Distribution and Ecology of
Living Benthic Foraminiferids

By the same author

An Atlas of British Recent Foraminiferids

Distribution and Ecology of Living Benthic Foraminiferids

JOHN W. MURRAY, A.R.C.S., B.Sc., D.I.C., Ph.D.

Lecturer in Geology,
University of Bristol

Distributed in the United States by
CRANE, RUSSAK & COMPANY, INC.
347 Madison Avenue
New York, New York 10017

 Heinemann Educational Books

Heinemann Educational Books

LONDON EDINBURGH MELBOURNE AUCKLAND TORONTO
HONG KONG SINGAPORE KUALA LUMPUR
IBADAN NAIROBI JOHANNESBURG NEW DELHI

ISBN 0 435 62431 8

© John W. Murray 1973

First published 1973

Published by Heinemann Educational Books Limited
48 Charles Street, London W1X 8AH

Printed in Great Britain
by Richard Clay (The Chaucer Press), Ltd
Bungay, Suffolk

Preface

The study of the distribution of living foraminiferids dates from the introduction in 1952 of Walton's method of staining protoplasm with rose Bengal. During the past twenty years, there has been a steadily increasing volume of published data based on the use of this staining method. In this book I have attempted to summarize and synthesize these data to produce a coherent pattern of foraminiferid distributions. The methods used are simple. I have avoided sophisticated computer interpretations because they are not readily understood, and because many micropalaeontologists are rightly suspicious of 'weighted' and 'adjusted' data.

The decision to restrict the book to living foraminiferids is based on the knowledge that, in many areas of the world, there are differences between the living and dead assemblages which are attributable to the Pleistocene fluctuations of sea level. A comparison of such a dead assemblage with present environmental conditions would be pointless and misleading. However, although no reference is made to their results, the pioneer works of d'Orbigny, Williamson, Brady, Heron-Allen and Earland, Cushman, and Natland should not pass unrecorded.

Carefully collected samples of adequate size are a prerequisite of ecological studies. With this in mind, an assemblage count of 100 individuals has been chosen as the smallest acceptable size for use in the numerical analyses. The methods used are described in Chapter 1. Then follows a series of chapters discussing the distribution of foraminiferids in the major environments. Chapters 17 to 20 consider essentially biological aspects, while Chapters 21 and 22 relate living to dead distributions and suggest a method of palaeoecological interpretation of fossil forms. The appendices give data on sampling methods and on the ecological requirements of genera.

Throughout the book I have attempted to use a consistent set of specific and generic names, and where these differ from those used by the original author, this is indicated in parentheses. In some cases these nomenclatural differences have been discussed with the author and some measure of agreement has been reached (e.g. Haake, Lutze).

In a synthesis of this kind, fields that require further investigation become apparent. To date, most emphasis has been placed on intertidal, shallow-water, nearshore and lagoonal areas. Much more information is needed on shelf, slope and deep seas, coral reefs and all tropical and arctic shallow water environments. There is also a need for long-term studies of small areas to attempt to relate the annual and diurnal variations in as many environmental parameters as possible to changes in the abundance and composition of foraminiferid assemblages. Preferably, these areas should be in shelf seas rather than in lagoons, marshes or other marginal marine situations.

It is a pleasure to record my thanks to all those who have helped to produce this book. Professor H. Sandon, Dr C. A. Wright and Dr J. Rogers have read chapters and made many useful suggestions for improvements. Mrs G. F. Murray carefully typed the manuscript. Mrs A. Gregory drafted the figures. Dr J. Rogers recomputed many of the published data. Mr C. J. Spittal and his staff gave invaluable assistance with library facilities. Prof. H. Hinton kindly allowed access to the scanning electron microscope.

1973   J.W.M.

Contents

CHAPTER 20 ECOLOGICAL CONTROLS

CHAPTER 21 RELATIONSHIP BETWEEN LIVING AND DEAD ASSEMBLAGES

List of Plates

1. Methods of Analysis

In this chapter particular emphasis is given to the methods used in this book, but other methods are briefly described also.

RELATIVE AND ABSOLUTE ABUNDANCE

The two methods of quantifying living animal assemblages are relative abundance, in which the number of individuals of each species forms a percentage or ratio of the total number of individuals (100 per cent), and absolute abundance, in which the number of individuals is related to a unit area or volume of sea floor.

The relationship between the two methods is shown in the following theoretical model, which has been designed to incorporate some of the features seen in natural assemblages. A profile of stations (A–L) shows a progressive increase in the absolute abundance and in the number of species present in the assemblage (see Figure 1 and Table 1). Comparison of the two methods of presenting the results shows:

    (a) Species 1, having a uniform absolute abundance, has a spurious peak at station A by the relative abundance method. Similar spurious peaks at station A are seen in the relative abundance of species 2, 4 and 5;

    (b) the peaks of abundance appear at different stations according to the method used (species 2, 3, 5 and 6);

    (c) most other species show a progressive increase in absolute abundance but this is not evident from the relative abundance histograms because the rate of increase of the abundance of individual species is roughly the same as that of the total population.

An additional disadvantage of the relative abundance method is that when the number of species is small (less than ten) the percentages of individual species are inevitably high and *vice versa*. Ujiié (1962) has pointed out that when the sampling error exceeds the percentage of a given species, the result should be rejected. He concluded that it was preferable to reject all occurrences of less than 7–10 per cent. While this might be desirable from the statistical point of view, it is scarcely practical.

Two definitions of absolute abundance have been used in the study of foraminiferids. Schott (1935) used the *Foraminiferal Number*, i.e. the number of foraminiferids present in 1 g dry weight of sediment. Phleger (1960a) and other authors have used the absolute abundance of foraminiferids in a unit area of 10 cm² (volume 10 cm³). Only the latter definition is appropriate to the study of living foraminiferids, but it has the disadvantage that the area of sea floor sampled must be known. Unless the rate of sedimentation and the

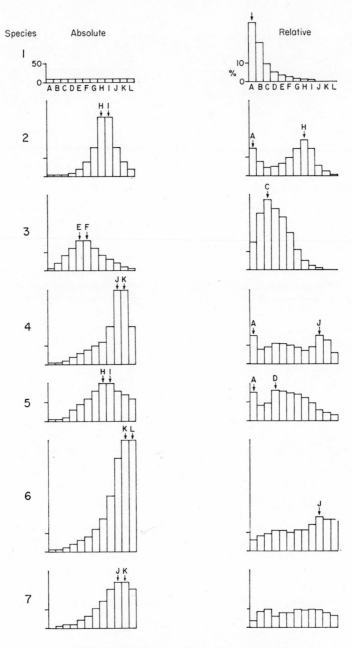

Figure 1 Histograms of absolute and relative
abundance of fifteen species along a profile
of stations A to L, based on the data in Table 1.

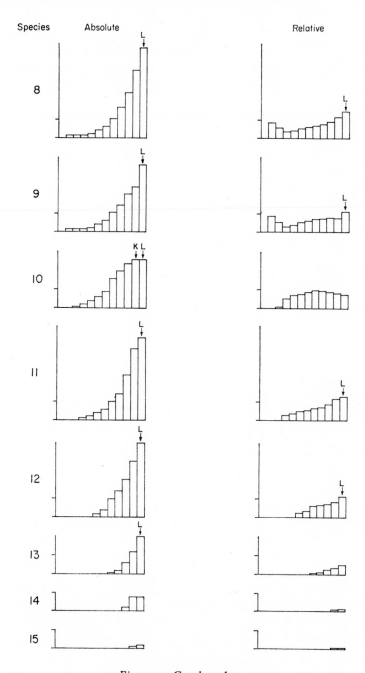

Figure 1—Continued.

Table 1 Theoretical model to show the relationship between relative and absolute abundance

Absolute abundance

Station	Species no. 1	2	3	4	5	6	7	8	9	10	11	12	13	14	15	Total
A	10	5	5	5	5	2	1	—	—	—	—	—	—	—	—	33
B	10	5	20	5	5	5	5	5	5	—	—	—	—	—	—	65
C	10	5	40	10	10	10	10	5	5	1	—	—	—	—	—	106
D	10	10	60	20	30	20	10	5	5	10	5	—	—	—	—	185
E	10	20	80	30	45	30	20	10	10	20	10	—	—	—	—	285
F	10	40	80	40	60	40	30	20	20	30	20	10	—	—	—	400
G	10	80	60	50	80	60	50	30	30	45	30	20	—	—	—	545
H	10	160	40	60	100	90	70	50	50	80	50	50	1	—	—	811
I	10	160	30	100	100	150	100	80	70	100	70	70	10	—	—	1050
J	10	80	20	200	80	250	120	120	100	120	120	100	30	10	—	1360
K	10	40	10	200	70	300	120	180	120	130	190	150	60	40	5	1625
L	10	20	5	100	60	300	100	240	180	130	220	200	100	40	10	1715

Table 1—Continued.

Relative abundance

Station	Species no. 1	2	3	4	5	6	7	8	9	10	11	12	13	14	15	Total
A	31	15	15	15	15	6	3	—	—	—	—	—	—	—	—	100
B	15	8	30	8	8	8	8	8	8	—	—	—	—	—	—	101
C	9.4	4.7	37.8	9.4	9.4	9.4	9.4	4.7	4.7	0.9	—	—	—	—	—	99.8
D	5.4	5.4	32.5	10.8	16.2	10.8	5.4	2.7	2.7	5.4	2.7	—	—	—	—	100.0
E	3.5	7.0	28.0	10.5	15.8	10.5	7.0	3.5	3.5	7.0	3.5	—	—	—	—	99.8
F	2.5	10.0	20.0	10.0	15.0	10.0	7.5	5.0	5.0	7.5	6.0	2.5	—	—	—	100.0
G	1.8	14.8	11.0	9.3	14.8	11.0	9.3	5.5	5.5	8.3	5.5	3.7	—	—	—	100.5
H	1.2	19.7	4.9	7.4	12.3	11.1	8.6	6.2	6.2	9.7	6.2	6.2	0.1	—	—	99.8
I	0.9	15.2	2.7	9.5	9.5	14.2	9.5	7.6	6.7	9.5	6.7	6.7	0.9	—	—	99.6
J	0.7	5.8	1.5	14.7	5.8	18.4	8.8	8.8	7.3	8.8	8.8	7.3	2.2	0.7	—	99.6
K	0.6	2.5	0.6	12.3	4.3	17.4	7.4	11.0	7.4	8.0	11.7	9.4	3.7	2.5	0.3	99.1
L	0.6	1.2	0.3	5.8	3.5	17.5	5.8	14.0	10.5	7.6	12.8	11.6	5.8	2.3	0.6	99.9

rate of production of the foraminiferids are known, there can be no direct comparison of living populations on a two-dimensional surface with Foraminiferal Numbers based on three-dimensional samples drawn from fossil assemblages.

Thus there are advantages and disadvantages of the relative and absolute abundance methods, but both are useful for different aspects of study as long as the user is aware of the errors which can arise.

STANDING CROP

The number of individuals present on a unit area of sea floor at any one time is the standing crop. Many authors use a unit area of 10 cm^2 but, even where a larger unit area is studied (e.g. Murray, 1968b, 1969), the results can be scaled down for comparative purposes. Standing crop can also be measured as biomass (i.e. live weight), dry weight, dry organic matter, displacement volume or calculated volume (see Nielson, 1963, p. 133) but these are not normally used in foraminiferid studies.

BIOMASS

Biomass is normally expressed as live weight, and can be used as a measure of standing crop. However, since a population of a few large animals may be

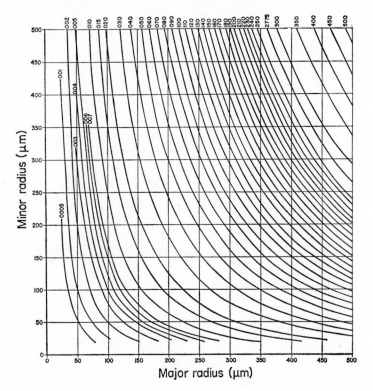

Figure 2 Size–volume graph of sphere, oblate and prolate sphaeroids. Volume contours are in decimal parts of a cubic millimetre, e.g. 002 = 0.002 mm^3.

ecologically more significant than a large population of small individuals, it seems worth using both measures of population size.

In the case of foraminiferids, it is difficult to measure live weight because of their small size and the difficulty of separating them from the sediment. It is therefore easier to express biomass as the calculated volume of the entire shell (Murray, 1968b) although in the case of large genera it is probably advisable to include only the volume of the chambers.

It is easier to approximate foraminiferids to simple geometrical shapes than to develop a formula to express the volume of a complex shape. Many foraminiferids are close to prolate or oblate sphaeroids, spheres or cones. The size-volume relationships are shown in Figures 2 and 3. It can be seen that as size increases arithmetically, volume increases logarithmically.

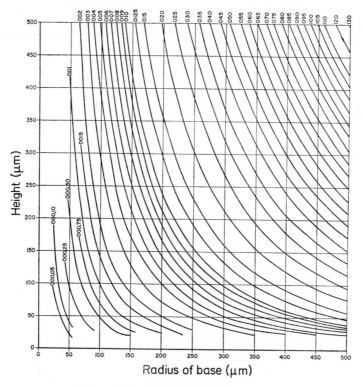

Figure 3 Size–volume graph of cone. Volume contours as in Figure 2.

Therefore, the most reliable way of measuring biomass is to measure each specimen and to determine its volume. However, in practice this is tedious, slow and not worth the labour involved. A more rapid method is to divide the individuals into a number of size groups and, by inspection, to select the middle-sized specimen in each group, to determine its size and volume, and to multiply by the number in the group. The sum of the group volumes will give the biomass.

DIVERSITY INDICES

Diversity is the relationship of the number of species to the number of individuals in an assemblage. If all assemblages comprised the same number of individuals, the numbers of species could be compared directly. However, normally it is necessary to compare assemblages of differing size. A variety of diversity indices have been proposed to enable this to be done. Sanders (1968) has given a comprehensive discussion of these indices. Here it is intended to describe only those diversity indices which have been used to interpret foraminiferid data.

Yule–Simpson index

Following Yule's (1944) study of literary vocabulary, Simpson (1949) modified one of his equations for use as a diversity index:

$$\frac{N(N-1)}{\sum_{i=1}^{K} n_i (n_i - 1)}$$

where N is the total number of individuals, K is the number of species and n_i is the number of individuals of the ith species.

The disadvantage of this index is that it is controlled mainly by the abundant species.

Gibson (1966) used this index to re-interpret data from Phleger (1954). Phleger made counts of 300 individuals and calculated the total fauna for each standard 10 cm³ sample. Gibson has re-interpreted these figures in such a way that the number of species is related to the total sample (which may be 4000 or more) rather than to the count of 300 individuals. This obviously is wrong. Ikeya (1971) used this index for total populations from North Japan.

Fisher α index

This index was first described by Fisher, Corbett and Williams (1943).

$$\alpha = \frac{n_1}{x}$$

where x is a constant having a value less than 1 (this can be read from Figure 125 of Williams, 1964) and n_1 can be calculated from $N(1 - x)$, N being the size of the population.

However, there is no need to calculate the index for each sample, as, once a base-graph has been constructed (Figure 4), the α value can be determined by plotting the number of species against the total number of individuals.

The α index takes the rarer species into account, and Williams (1964) considers that it is a consistent index where the distribution is assumed to be of the log-series form. Murray (1968b) made successively larger counts on three samples to check the constancy of α. The variation was shown not to

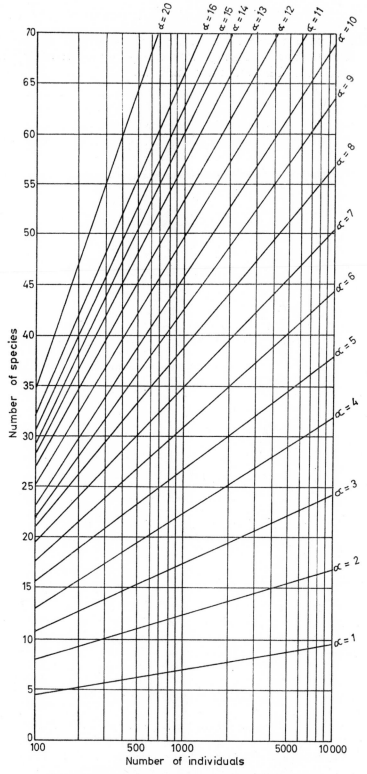

Figure 4 Graph to show the relationship between the number of species, the number of individuals in an assemblage and lines of equal α diversity index.

be great, but there was a tendency for α to increase with sample size (Figure 5). Nevertheless, as this diversity index is very easy to use and produces useful results, it has been adopted in this book.

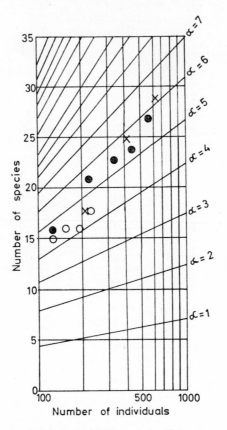

Figure 5 Variation of α index according to sample size (after Murray, 1968b).

Information function

Buzas and Gibson (1969) used the information function to measure foraminiferal diversity. They pointed out that it '... measures the number of species and their proportions without making any assumptions as to an underlying distribution'. The function is

$$H(S) = \sum_{i=1}^{S} p_i \log p_i$$

where p_i is the proportion of the ith species, and S is the number of species.

Maximum diversity, i.e. $H(S)$, is reached when all species have equal frequencies. Buzas and Gibson plotted $H(S)$ against depth for foraminiferal assemblages from the continental shelf and ocean off the east coast of the U.S.A. Diversity peaks were found at 35–45 m, 100–200 m and deeper than

2500 m. They also plotted a measure of species equitability, which was the ratio

$$e^{H(S)}/S$$

where e is the base of the natural logarithms. Species which are perfectly equally distributed have the ratio equal to 1.

The use of information theory in diversity studies has also been discussed by Beerbower and Jordan (1969).

Cumulative curves

A graphical approach to diversity is to plot the logarithm of the number of individuals against the number of species as a cumulative curve (Odum, Cantlon and Kornicker, 1960; Williams, 1964; Shaffer, 1965). Low-diversity samples plot low on the graph and have a gentle slope, whereas high diversity samples plot high on the graph and have a steep slope. Since lines of equal α value can be plotted on the same graph, the relationship between the two measures of diversity is clearly seen (Figure 6). The difference in interpreta-

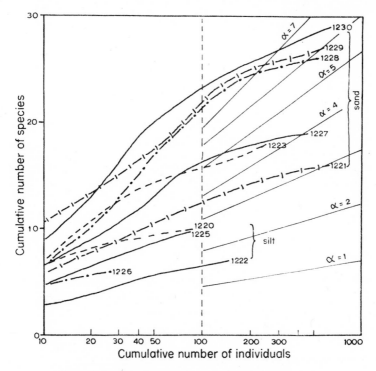

Figure 6 Cumulative curves of Buzzards Bay samples (after Murray, 1968b).

tion needs to be stressed; for the cumulative curves the position of the curve and its angle of slope are important, whereas for α it is the upper end-point of the curve which defines the index.

When using cumulative curves, it is unwise to use percentage data (as in Walton, 1964a) as the effects of sample size are eliminated and, unless the samples were all originally of the same number of individuals, errors of interpretation may result. Extensive use of the method was made by Murray (1968b). However, it is a time-consuming method of plotting diversity, and has therefore been abandoned in favour of the α index.

COMPARING SAMPLES: SIMILARITIES AND DIFFERENCES

Sanders (1960) described a method of measuring similarity between samples in order to construct a trellis diagram. The percentage occurrence of species in the two samples to be compared are listed side by side. For each species common to the two samples, the lowest percentage occurrence is taken and the total value is the similarity index. On average, values higher than 80 per cent indicate that the samples are nearly identical. Lower values indicate progressively greater differences. For example:

Species	#1	#2	% in common
A	10	10	10
B	20	28	20
C	15	7	7
D	15	16	15
E	5	—	—
F	30	25	25
G	2	—	—
H	3	14	3
	100	100	80% similarity

Murray (1969) has used similarity indices to compare adjacent samples along traverses. They can also be used to compare adjacent samples areally or to compare all samples from one area. For the latter it is advisable to make use of a computer to reduce the calculation time.

TRIANGULAR PLOT OF SUBORDERS

It is convenient that all modern foraminiferids with hard tests fall into three suborders, the Textulariina, Miliolina and Rotaliina (Loeblich and Tappan, 1964). They therefore lend themselves to plotting on a triangular diagram (Figure 7).

SPECIES DOMINANCE

Walton (1964a) introduced the concept of *faunal dominance* which he defined as '... the percentage occurrence of the most common species in a foraminiferal population'. In general, very variable environments are dominated by few species which have high abundance (90–100 per cent in some marshes), whereas stable environments are characterized by many species, none of which is very abundant (10–15 per cent in shelf seas).

To a certain extent, species dominance is controlled by the number of species in an assemblage; the more species present, the lower is the percentage occurrence of each. Walton prepared a graph based on data from the Gulf of Mexico. In samples of less than 20 species, the dominant species formed at least 25 per cent. In samples of more than 20 species, the dominant species normally formed less than 35 per cent.

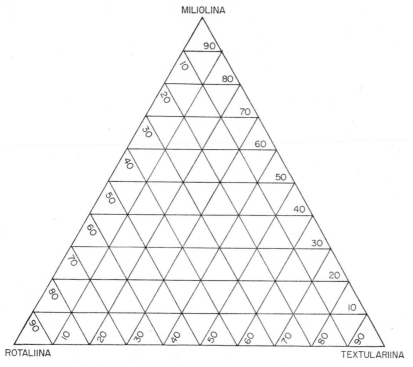

Figure 7 Triangular diagram of the ratio of the three suborders.

Another way of looking at species dominance is to use the entropy concept. This was first described by Pelto (1954) and used to interpret foraminiferid data by Miller and Khan (1962) and Howarth and Murray (1969). Entropy (*H*) is derived from information theory:

$$H = - \sum_{i=1}^{n} p_i \log p_i$$

In practice the relative entropy (*H_r*) is used:

$$H_r = \frac{100H}{H_m}$$

where the maximum value of *H* is H_m.

Miller and Khan give a list of values for $-p \log p$ and proportions (*p*; this is percentage divided by 100; 10 per cent = 0.10*p*) and also a table of H_m values for up to 10 components (1962, p. 428).

Relative entropy is at a minimum when the composition is 100 per cent of one species, and at a maximum where the contribution of all species is all the same. A map of H_r values should ideally have zones of low entropy (i.e. single-species dominance) separated by zones of relatively high entropy in which there are several dominance species. Examples of such images are given by Miller and Khan (1962, p. 431) and Howarth and Murray (1969, p. 672). It is important to realize that although low entropy indicates that the assemblage is close to 100 per cent of one species, not all low entropy areas are related to the same species.

COMPUTER STUDIES

Apart from the general use of computers to do routine calculations on foraminiferid assemblage data, there have been two main fields of application: factor analysis and cluster analysis.

Factor analysis was used by Howarth and Murray (1969) in a re-appraisal of the ecology of Christchurch Harbour, England. Cluster analysis has been more commonly used: Kaesler (1966), on Todos Santos Bay; Mello and Buzas (1968), on the central Texas coast; Howarth and Murray (1969), on Christchurch Harbour, England; and Ujiié and Kusukawa (1969), on Japanese bays.

The methods of computation are complex, and for details of procedure the reader is referred to the above papers.

LIVE-DEAD RATIO

Walton (1955) introduced the ratio

$$\frac{L}{D} = \frac{\text{number of living}}{\text{number of dead}} \times 100$$

as a means of assessing the rate of sedimentation. Use of this ratio has become quite common. However, few authors have considered the errors involved.

Where seasonal studies of standing crop size have been carried out, the annual range has often been ten times the lowest value, e.g. in 1965 the variation for *Elphidium articulatum* (d'Orbigny) in Bottsand Lagoon is from less than 10 in January to 125 in August (data from Lutze, 1968a). A conservative estimate would be a threefold difference. Suppose a dead population of 1000 individuals and standing crops of 100 and 300. The resulting L/D ratios would be 10 per cent and 30 per cent respectively. Boltovskoy and Lena (1969b) calculated L/D ratios for 24 monthly samples from the same locality in Puerto Deseado, Argentina. The range of variation was 5.6 to 50.0 per cent.

The only reliable method of assessing the rate of sedimentation from foraminiferids is to compare the annual production with the number of dead individuals (Murray, 1967b).

INDEX OF REGENERATION

Boltovskoy (1957) noted that irregularities in the tests of foraminiferids could have two possible origins, ecological or mechanical. In the latter case the animal repairs its test. In material from the Rio de la Plata, *Elphidium* had the best powers of regeneration, 0.08 per cent of the assemblage showed signs of regeneration and Boltovskoy proposed this measure as a *regeneration index*. Although he postulated that different values would be obtained from different environments, this index has not been used subsequently.

SALINITY

Throughout this book the following terms have been used: hyposaline = <32 per mille, normal marine = 32 — 37 per mille, hypersaline = >37 per mille. The subdivisions of hyposaline (brackish) waters, such as those listed by Hiltermann (1948), have not been followed.

2. *The Water's Edge*

Because the sea is subject to tidal movements and to the effects of periodic storms, the meeting of sea and land is not a single line but a zone of varying width, the sea shore. To most people the sea shore is synonymous with sandy beach but this is only one aspect of shores. Other types of shoreline are mudflats, marshes, mangrove swamps and rocky coastlines.

From the ecological point of view, the interest in shorelines is that fauna and flora of the sea meet those of the land. However, whereas some land plants have developed the ability to survive submergence by the sea, and some seaweeds to survive exposure to the atmosphere, in the case of animals the indigenous fauna of the shore zone is almost entirely of marine origin.

To the geologist, the interest in shore-zone sediments is that they usually herald a marine transgression or terminate a phase of marine sedimentation. Therefore the recognition of these environments from their animal assemblages is important.

The area between high water and low water is commonly termed the littoral or intertidal zone. The exact seaward and landward limits are, of course, variable because they depend on the tidal amplitude variation between spring and neap tides. Apart from the alternate flooding and ebbing of the tides, the most important feature of the intertidal zone is the effect of waves. In situations where waves expend a large amount of mechanical energy, the coast can be spoken of as a high-energy shoreline. Where waves expend very little mechanical energy, there is a low-energy shoreline.

A simple classification of shoreline types is:

> *Low energy*
> Tidal marsh and mangrove swamp
> Mudflats
>
> *High energy*
> Sandy beaches
> Rocky shorelines (including cliffs)

Whereas the factors controlling high-energy shorelines are usually regional, e.g. the occurrence of open waters with large fetch for the waves, the factors controlling low-energy coastlines are sometimes local, e.g. marshes and mudflats developed around the shores of a lagoon or estuary protected from the open sea by a sand spit or barrier inland.

LOW ENERGY

TIDAL MARSHES

Marshes are tidal flat areas covered by vegetation. They are known also as salt marshes or saltings. This environment is of great interest to ecologists because it represents a truly transitional region between either a marine or a brackish aqueous environment and a terrestrial environment. Representatives of the former environments include diatoms, other microscopic algae, *Enteromorpha*, crabs and bivalves, while those of the latter environment include *Spartina*, *Salicornia* and many other phanerogam halophytes, together with birds, mosquitoes, etc. Tidal marshes are typical of the temperate regions; in the tropics the comparable environment is a mangrove swamp.

The following features are characteristic of a tidal marsh:

(a) It is developed in the upper part of the intertidal zone.
(b) During much of each tidal cycle it is exposed to the atmosphere.
(c) It spends a smaller part of the tidal cycle submerged beneath sea or estuarine water.
(d) It is subject to the more dramatic fluctuations in temperature of the atmosphere.
(e) Fresh water is periodically introduced from rain showers.
(f) There is usually high organic productivity by the marsh plants and microflora.
(g) The fauna comprises few permanent indigenous species, although migratory terrestrial animals are sometimes common.

Marshes are really a complex of environments. The seaward part usually spends longer submerged under water than does the landward part. This is reflected by floral differences; *Spartina* is usually dominant seawards, *Salicornia* landwards, and other plants such as *Batis* and *Distichlis* are found on the highest marsh. All these plants are essentially land plants which have invaded the upper part of the intertidal zone. Marsh floras have been reviewed by Hedgpeth (1957b).

The marsh surface is rarely absolutely smooth. There are local highs representing former channel levees, and shallow pools (sometimes termed salt pans, see Evans, 1965) representing unfilled portions of former channels. The pools vary greatly in temperature and salinity according to their position on the marsh and the prevailing weather conditions. In dry periods they may dry up to leave a thin salt crust. At other times rainfall may cause a rapid lowering of temperature and salinity. In some marshes small erosional clifflines are developed separating portions of marsh of different elevation. Most commonly, a cliff separates the higher marsh with a *Salicornia* flora from the lower marsh with *Spartina*.

A most comprehensive analysis of the environmental parameters of a marsh has been carried out by Bradshaw (1968). The area studied was a nature reserve on the north shore of Mission Bay, San Diego, California. Measurements were made from January 1964 to August 1965. An example of the diurnal rhythm at one station is shown in Figure 8 (Phleger and Bradshaw, 1966).

The flood tide introduces water of salinity 34 per mille and pH 8.0–8.2 on to the marsh. When this water ebbs, it is seen to have a higher salinity (up to 50 per mille), a more variable pH (8.5 maximum during the day and 6.8 minimum at night) and a more variable temperature. Seasonal changes

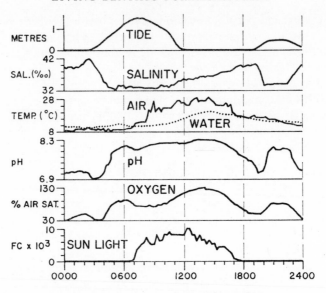

Figure 8 Diurnal rhythm in Mission Bay tidal marsh
(redrawn from Phleger and Bradshaw, 1966).

were not obvious for pH and oxygen concentration but they were clearly
shown in the temperature of the water. The mean monthly temperature
varied from 12.7 °C in December and January to 24.1 °C in July. However,
the temperature range for any one month often exceeded 10 °C. The coldest
time was December (minimum 4.9 °C) and the warmest was July (maximum
33.0 °C). Salinity data were available only for September and October. Therefore no long seasonal changes were observed, although the mean monthly
values differed (37.2 per mille and 39.0 per mille respectively).

Mission Bay experiences the dry climate of California. The marsh receives no land runoff and from this point of view, perhaps, cannot be regarded as typical. However, these detailed observations confirm the generally held view that the salt marsh is an environment of extreme variation.

Within the marsh complex, certain microhabitats may be more favourable than others. The larger plants shade the marsh beneath and thus protect it from extreme temperature changes. Algal mats keep the sediment
moist. As the sediment is normally anaerobic a few millimetres below the
surface, foraminiferids are restricted to the superficial oxidized layer and to
the vegetation. Their food is supplied by the marsh plants, filamentous algae,
diatoms and other small algae. The subject has been reviewed by Murray
(1971b).

Hyposaline marshes In North America the following studies have been
made:

1. Poponesset Bay, Massachusetts, Parker and Athearn (1959), Rappahannock River, Virginia, Ellison and Nichols (1970).
2. Galveston Bay, Texas, Phleger (1965a).

3. Pacific coast—Fraser River Delta, Gray's Harbour, Coos Bay, Phleger (1967).

4. Estero Punta Banda, Baja California, Walton (1955).

The studies which have been made in Europe are:

1. Dovey Estuary, Wales, Haynes and Dobson (1969).
2. Bottsand Lagoon, Baltic, Lutze (1968a).
3. South Holland (polyhaline parts), Phleger (1970).
4. Arcachon, Le Campion (1970).

The known environmental parameters are summarized in Table 2.

Table 2 Summary of the known environmental parameters
of hyposaline tidal marshes

Locality	Salinity (per mille)	Temperature (°C)	Tidal range (m)
Massachusetts			
Poponesset Bay	0.6–34.3*	0.2–30	not quoted
Gulf of Mexico			
Galveston Bay	av. 30	27–28 summer	0.3–0.75
	av. 22	2–27 winter	
Pacific Coast			
Fraser River Delta	not quoted	not quoted	not quoted
Gray's Harbour	not quoted	not quoted	not quoted
Coos Bay	not quoted	not quoted	not quoted
Estero Punta Banda	34–37	15–30	1
Europe			
Dovey Estuary	17–36	0–25	2.13–4.88
Severn Estuary	not known	0–26	5–14
Bottsand Lagoon	6–14	0–30	<1
	rarely up to 40		
South Holland	polyhaline mesohaline	not quoted	not quoted
Arcachon	4–6	2–25	2–4.5

* 1 value of 45.2 per mille on isolated high marsh pool.

For all these marshes the diversity is very low, α values less than 3 (Figure 9). The triangular plot shows the general absence of Miliolina (except in South Holland polyhaline marshes) and dominance of Textulariina and Rotaliina in the majority of assemblages (Figure 10). Details for the individual marshes are listed in Table 3. The standing crop is highly variable, with values of 0 to 8600 individuals per 10 cm². The occurrence of individual species is listed in Table 5, page 25. Only *Miliammina fusca* (Brady) is common to all hyposaline marshes. *Ammotium salsum* (Cushman and Brönnimann), *Trochammina inflata* (Montagu) and *Jadammina macrescens* (Brady) are fairly widespread.

Adjacent to Bristol is the Severn estuary, a region noted for its powerful currents and extreme tidal range (5 m neap, 14 m spring). *Spartina* marshes near Bristol are submerged by only the higher of the high tides, some parts

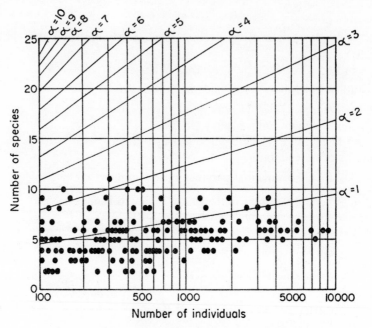

Figure 9 Plot of the α diversity values for hyposaline marshes (Poponesset Bay, Galveston Bay, Fraser River Delta, Gray's Harbour, Coos Bay, Dovey Estuary, South Holland).

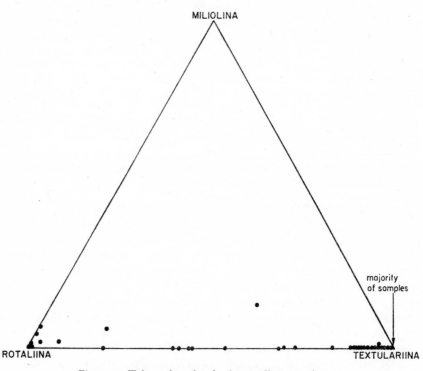

Figure 10 Triangular plot for hyposaline marshes (Poponesset Bay, Galveston Bay, Fraser River Delta, Gray's Harbour, Coos Bay, South Holland).

Table 3 Summary of some characteristics of the foraminiferids
of tidal marshes

	α index	Standing crop per 10 cm²	% Textulariina	% Miliolina
Hyposaline				
Poponesset Bay	1.5–2.5	90–400	41–97	0
Galveston Bay	0–2	0–1300 often 100	9–100	0
Fraser River Delta	<1	430–7250	100	0
Gray's Harbour	0–2	25–5600	99–100	0
Coos Bay	<1.5	51–8600	100	0
Estero de Punta Banda	<1	0–108	96	0
Dovey Estuary	0–2	0–700	5–60	0
Bottsand Lagoon	1	0–261	22–98	0
South Holland	1	0–1320	0–99	0–13
Arcachon	0–2.5	—	95–100	<1
Normal marine				
Mission Bay	<1.5	2–4400 often <1000	32–100	0–51
Norfolk	<1.5	1–778	23–100	0–6
South Holland	<1.5	4–1200	0–94	0–21
Hypersaline				
South Texas	0–4	10–3300 (see Table 4)	0–100	0–89
Guerrero Negro	0–5	0–800	5–100	0–73
Ojo de Liebre	1	0–117	36–80	19–61

are wetted only fifty or so times a year and then just for one to two hours.
During the summer the marsh dries, during the winter it freezes, and
throughout the year it is subject to periodic rainfall. It is probably one of the
most extreme marsh environments known. Living foraminiferids are extremely rare and when present usually one species, *Elphidium articulatum*
(d'Orbigny), makes up the entire assemblage. This environment is close to
the limits for the existence of foraminiferids. This is reflected in the extremely low diversity values (<1).

Normal marine marshes   The best known example is Mission Bay, California (Bradshaw, 1968; Phleger, 1967; Phleger and Bradshaw, 1966).
The flood tide introduces water of 34 per mille salinity and pH 8.0 to 8.2 on
to the marsh. On the ebb, the water is seen to have a higher salinity (up to
50 per mille), a more variable pH (8.5 maximum during the day and 6.8
minimum at night) and a more variable temperature. Seasonal temperature
data showed a monthly average of 12.7 °C in December and January and
24.1 °C in July. The minimum was 4.9 °C in December and the maximum
33.0 °C in July. The climate is dry and there is no land runoff. Tides are
small (1–1.5 m). Two less well-known examples are Scolt Head, Norfolk,
and the euhaline marshes of South Holland (Phleger, 1970).

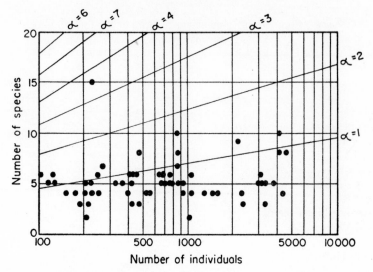

Figure 11 Plot of the α diversity values for normal marine marshes (Mission Bay, California, Norfolk, South Holland).

The diversity of the foraminiferid assemblages is low α = <2 (with one value of 3.5, Figure 11). The triangular plot shows a dominance of Textulariina with subsidiary Miliolina and Rotaliina (Figure 12). The dominant

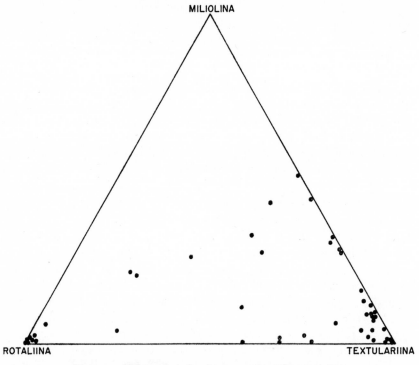

Figure 12 Triangular plot for normal marine marshes (Mission Bay, California, Norfolk, South Holland).

foraminiferids are *Jadammina polystoma* Bartenstein and Brand, *Miliam-mina fusca* (Brady) and *Trochammina inflata* (Montagu) but each species shows a varying pattern of distribution. In the Mission Bay marsh, *J. polystoma* and *T. inflata* characterize areas of *Salicornia* and are rare in channels. *M. fusca* is rare on tidal flats, marsh pools and channels. Miliolids and *Ammonia beccarii* (Linné) are found only in the marsh channels. The standing crop is generally large (total range 2 to 4400 individuals per 10 cm²), with the highest values in the *Salicornia* zone (generally > 1000 per 10 cm²).

Hypersaline marshes Most of the available data concern the South Texas coastal lagoon marshes (Phleger, 1966a) but there are also data from Baja California (Phleger and Ewing, 1962; Phleger, 1967).

The South Texas marshes were sampled in August 1963, following a period of drought. Although no salinity or temperature data were recorded, Phleger considered the environment to be hypersaline. Tides are small (0.3–0.6 m). Altogether, eleven different marsh areas were sampled from Matagorda, San Antonio, Aransas, Corpus Christi and Copano Bays. The main floral elements are *Spartina* on the lower marsh and *Salicornia* at higher levels.

In Baja California, Guerrero Negro marsh is slightly hypersaline (salinity 35.5–37.5 per mille in July) while nearby Laguna Ojo de Liebre marsh has salinities of 42 to greater than 60 per mille.

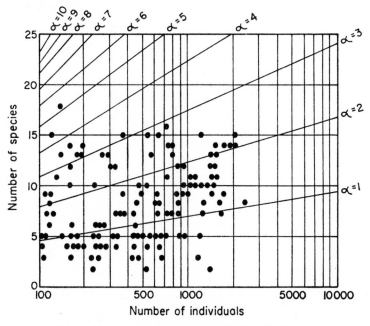

Figure 13 Plot of the α diversity values for hypersaline marshes (South Texas lagoons, Guerrero Negro, Ojo de Liebre).

Figure 13 shows 109 marsh assemblages and 39 adjacent channel and bay assemblages from the Texas lagoons, and 22 from the Baja California marshes. The α values are α = < 4 for most assemblages with two values at

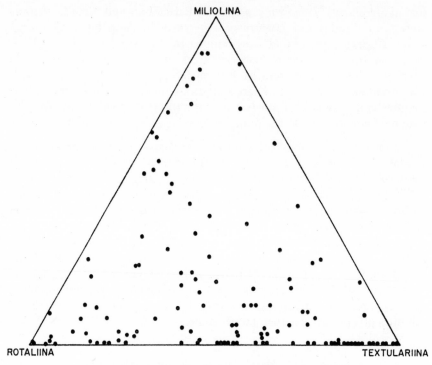

Figure 14 Triangular plot for hypersaline marshes (South Texas lagoons, Guerrero Negro, Ojo de Liebre).

Table 4 Standing crop per 10 cm² for south Texas marsh samples (after Phleger, 1966)

Zone	Number of samples	Range in size	Average size
Bay	28	185–2200	1400
Mud Flat	45	15–2250	935
Spartina alterniflora	49	25–1500	565
Spartina-Salicornia	12	25–540	240
Salicornia	18	25–530	235

4.5 and 5.5. On the triangular plot (Figure 14) the points fill the entire field. For the South Texas marshes, Phleger noted a higher dominance of Textulariina than in the adjacent channels and bay. The standing crop data are summarized in Table 4. Much smaller values are found in Guerrero Negro marsh (0–300 per 10 cm² and mainly less than 100), and in Ojo de Liebre marsh (0–4 per 10 cm² in brine pan, 30–187 per 10 cm² on *Salicornia*-covered levees). The species present are listed in Table 5. *Ammonia beccarii* (Linné) and *Miliammina fusca* (Brady) are common throughout the South Texas marshes (although the latter species is rare in the *Salicornia* zone).

Table 5 The occurrence of common species in marshes

	Poponesset Bay	Rappahannock	Galveston Bay	Frazer River delta	Gray's Harbour	Coos Bay	Estero de Punta Banda	Dovey Estuary	Bottsand Lagoon	South Holland	Arcachon	Mission Bay	Norfolk, England	South Holland	South Texas	Guerrero Negro	Ojo de Liebre
Ammotium salsum	X	X	X	X				X							H		
Arenoparrella mexicana	X	X	X									X			H		
Miliammina fusca	X	X	X	X	X	X	X	X	X	X	X	N	N	N	H		H
Protelphidium tisburyensis	X																
Tiphotrocha comprimata	X	X	X					X									
Ammonia beccarii		X	X									X	N	N	H		
Trochammina inflata		X	X		X	X		X	X	X	X	N	N	N	H	H	H
Ammobaculites sp.	X			X													
Pseudoclavulina sp.				X													
Jadammina macrescens	X				X	X		X	X						H		
Jadammina polystoma					X	X			X	X	X	N	N	N	H		
Haplophragmoides sp.	X			X				X									
'Proteonina' sp.							X										
Haplophragmoides subinvolutum							X										
Protelphidium anglicum*							X		X				N	N			
Elphidium articulatum1†							X	X			X						
Miliolinella spp.												N					
Quinqueloculina laevigata												N					
Miliolids									X				N	N	H	H	H
Elphidium spp.									X				N	N	H	H	
Pseudoeponides andersoni															H		
Discorinopsis aguayoi																H	
Glabratella sp.																H	
Glomospira sp.																H	
Textularia earlandi																H	

X—hyposaline marsh. N—normal marine marsh. H—hypersaline marsh.

   Protelphidium anglicum—P. depressulum of Haynes and Dobson, 1969, and *N. tisburyensis* of Phleger, 1970.

   †*Elphidium articulatum—Cribrononion articulatum* of Lutze, 1968a, and *E. excavatum* of Haynes and Dobson, 1969, and Le Campion, 1970.

Bay, mudflat and *Spartina alterniflora* marsh have abundant *Ammotium salsum* (Cushman and Brönnimann), miliolids and *Elphidium* spp. Phleger regards these species as lagoonal, and present in the marsh only because of its proximity and narrowness. The *Salicornia* and *Salicornia-Spartina* marsh are characterized by *Arenoparrella mexicana* (Kornfeld), *Jadammina polystoma* Bartenstein and Brand, *Pseudoeponides andersoni* Warren, *Tiphotrocha comprimata* (Cushman and Brönnimann), *Trochammina inflata* (Montagu), and *Jadammina macrescens* (Brady) (=*Trochammina macrescens* of Phleger).

Summary of living assemblages   The feature which characterizes the foraminiferid assemblages of all tidal marshes is the low diversity. In hyposaline and the normal marine marshes the α value is generally less than 2. In the hypersaline marshes it rises to values greater than 3, with rare values of 4 and 5 (see Table 3). Standing crop varies from 0 to 8600 per 10 cm^2 but no consistent pattern is observable, even within one marsh. Both these characteristics can be attributed to the very variable nature of the environment.

The triangular plots of the three suborders show marked differences between the three types of marsh. The fields have been summarized in Figure 15. Hyposaline marshes are characterized by the general absence of Miliolina and the dominance of Textulariina and Rotaliina (the few samples

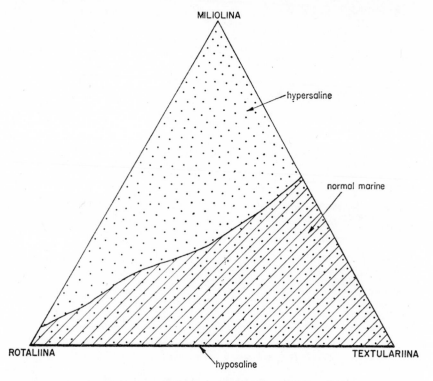

Figure 15 Summary triangular plot showing the fields for hyposaline, normal marine and hypersaline marshes.

with miliolids have been omitted). The normal marine marsh shows dominance of Textulariina with subsidiary Miliolina and Rotaliina. Hypersaline marshes span the entire triangle with mixtures or dominance of all three end-members.

The occurrence of the abundant and dominant species (Table 5) shows the following forms to be cosmopolitan with respect to latitude and type of marsh:

> *Ammotium salsum* (Cushman and Brönnimann)
> *Arenoparrella mexicana* (Kornfeld)
> *Miliammina fusca* (Brady)
> *Trochammina inflata* (Montagu)
> *Jadammina macrescens* (Brady)
> *Jadammina polystoma* (Bartenstein and Brand)

These can be regarded as the 'typical' marsh species. A second group is confined to hyposaline marshes:

Protelphidium tisburyensis (Butcher) ⎱ These may be synonymous
Protelphidium anglicum Murray ⎰
Tiphotrocha comprimata (Cushman and Brönnimann)
Ammobaculites sp.
Pseudoclavulina sp.
Haplophragmoides spp.
'*Proteonina*' sp.
Elphidium articulatum (d'Orbigny)

A third group of species is confined to hypersaline marshes:

Pseudoeponides andersoni Warren
Discorinopsis aguayoi (Bermudez)
Glabratella sp.
Glomospira sp.
Textularia earlandi Parker

Representatives of the second and third groups, together with *M. fusca* from the first group, are also found in non-marsh environments.

The miliolids found in normal marine and hypersaline marshes are simple, small species of *Quinqueloculina* and *Miliolinella*.

Hypersaline and normal marine marshes are not clearly separable (Mission Bay should perhaps be regarded as hypersaline anyway) but they are distinct from hyposaline marshes.

Geographical provinces In view of the environmental differences between hyposaline, normal marine and hypersaline marshes, it is desirable that the geographic distribution of species in each should be considered separately. (Phleger, 1967, discussed this topic but did not make such a separation). Since there are few examples of normal marine and hypersaline marshes, there is insufficient information to draw any reliable conclusions. There are more data on hyposaline marshes. On the Pacific Coast of North America, Phleger recognized subarctic Colombian and Aleutian assemblages with *Trochammina inflata* (Montagu) variant and *Miliammina fusca* (Brady) respectively. A second assemblage (Fraser River Delta, Gray's Harbour and Coos Bay) comprises northern species of *Elphidium*, *Jadammina polystoma* Bartenstein and Brand, *Miliammina fusca* (Brady), *Trochammina inflata* (Montagu), *Jadammina macrescens* (Brady), *Haplophragmoides subinvolutum* Cushman and McCulloch, *Ammotium salsum* (Cushman and Brönnimann) and *Ammobaculites exiguus* (Cushman and Brönnimann). A third assemblage in Galveston Bay, Gulf of Mexico, includes the following species which have not been recorded from the Pacific coast: *Ammoastuta inepta* (Cushman and McCulloch), *Arenoparrella mexicana* (Kornfeld) and *Tiphotrocha comprimata* (Cushman and Brönnimann). North American species not yet recorded from Western Europe include *Ammoastuta inepta* (Cushman and McCulloch) and *Arenoparrella mexicana* (Kornfeld).

Postmortem changes Several authors have noted that although calcareous shelled foraminiferids may be present in the living assemblages of marshes, they are much less commonly preserved in the sediments (Parker and Athearn, 1959; Phleger, 1966a; Bradshaw, 1968). Destruction of the tests is attributed to the reducing conditions which prevail beneath the superficial oxidized sediment layer. Observations by Bradshaw (1968) of pH changes show that if pH 7.6 is regarded as the equilibrium value for carbonate in sea water, the Mission Bay marsh is below this for approximately half the time. Living foraminiferids with calcareous tests could combat solution, but after death solution would be rapid. Bradshaw considered that it would take only one day at pH 5.0. Even if complete solution does not occur, the tests would be etched and weakened (Murray, 1967a, Murray and Wright, 1970).

Loss of calcareous species would cause a lowering of the α value of the dead assemblage and a move to the Textulariina corner on the triangular plot. Where complete solution takes place, it will not be possible to differentiate hyposaline from hypersaline assemblages. The fauna would still be recognizable as of marsh origin because of the presence of simple species of *Trochammina, Jadammina, Haplophragmoides* and *Ammobaculites*, and the very low diversity. Any such fossil assemblage should be suspected of being of marsh origin. If *Miliammina* is present the assemblage could represent a hyposaline lagoon. If there are common *Saccammina, Cribrostomoides, Reophax, Eggerella* and *Textularia*, the assemblage is more likely to represent a decalcified shelf assemblage.

MANGROVE SWAMPS

Mangroves characterize the marshes of tropical areas. The pioneer flora is sometimes the red mangrove (*Rhizophora*), followed landwards by the black mangrove (*Avicennia*), but elsewhere the black mangrove is the pioneer plant (Louisiana coast west of the Mississippi Delta, Hedgpeth, 1957b; Trucial coast lagoons, Evans *et al.*, 1964). Mangrove swamps develop along the borders of estuaries, along open marine coasts and in hypersaline environments.

On the Trucial Coast of the Persian Gulf, Murray (1965a, 1970a) has recorded infrequent *Ammonia beccarii* (Linné) in pools associated with mangrove areas. It seems likely that few foraminiferids can withstand the desiccation of subaerial exposure or the high salinities of tidal pools.

Phleger (1966b) described the foraminiferids in mangrove bay areas of south-western Florida. The area comprised Ten Thousand Islands, Whitewater Bay and part of Florida Bay. Subenvironments include mangrove swamps, islands and channels. Depths are typically less than 2–3 m. Freshwater runoff causes a salinity decrease from normal, open Gulf to 5 per mille by land during the summer months. In the winter higher salinities prevail (32–36 per mille). Temperatures are believed to range from 5–32 °C during the year.

In the Ten Thousand Islands area, the sediments are calcareous quartz sands and silts (20–30 per cent $CaCO_3$); in Whitewater Bay, quartz is uncommon and calcareous sands and silts prevail (60–80 per cent $CaCO_3$).

Standing crop ranges from 8–1489 per 10 cm² in Whitewater Bay, 72–1195 in the Ten Thousand Islands area, and 118–494 in Florida Bay. Variability is a characteristic feature.

Most of the assemblages are dominated by *Ammonia beccarii* (Linné) and various species of *Elphidium*. Locally *Ammotium salsum* (Cushman and Brönnimann) (=*Ammobaculites salsus* of Phleger) is common.

Avicennia mangrove marshes in New Zealand were studied by Phleger (1970). No environmental data were given. The standing crop values range from 0 to 2090 per 10 cm² but only 9 samples out of 63 had more than 100 individuals. The dominant species are patchily distributed. They include *Trochammina inflata* (Montagu), *Ammonia beccarii* (Linné) and *Haplophragmoides* sp.

Diversity ranges from α values of less than one up to 5.5, with higher values in the Florida marshes (Figure 16). On the triangular plot (Figure 17)

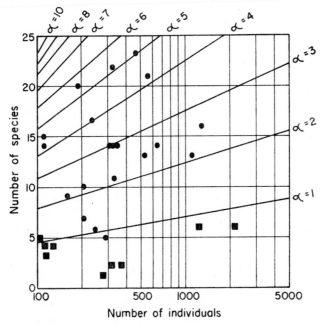

• Florida
■ New Zealand

Figure 16 Plot of the α diversity values for mangrove swamps.

the points are widely scattered. The Florida marshes never have more than 50 per cent Textulariina and they often have an important Miliolina component. The New Zealand examples lack Miliolina. The salinity of the Florida marshes is normal; that of the New Zealand ones is unknown.

TIDAL FLATS

Extensive areas of intertidal sediment are commonly called tidal flats. They are best developed in regions subject to large tidal ranges and particularly along shorelines which are protected from intense wave disturbance. They are common in large embayments (e.g. the Wash, England), in the protection of barrier islands (e.g. the Wadden Sea, Netherlands) and in estuaries (e.g. the Severn estuary, England).

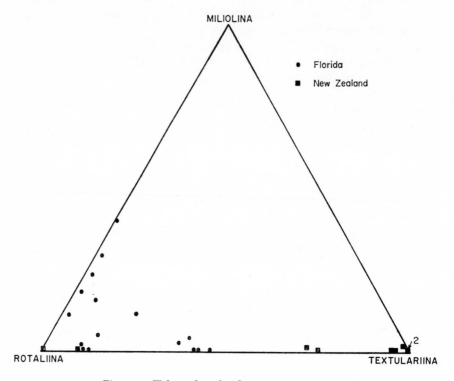

Figure 17 Triangular plot for mangrove swamps.

In areas of clastic sedimentation the sediment may be sand, mud or mixtures of the two. There is normally a gentle slope towards the sea or towards the tidal channels and creeks that intersect the tidal flats.

The macrofauna is mainly an infauna, but some forms spend part of their life on the sediment surface (see Evans, 1965). The fauna of a tidal flat must be able to tolerate the twice-daily influx and egress of sea water and the intervening periods of subaerial exposure. There are large diurnal and seasonal temperature changes, and the effects of evaporation or rainfall can drastically alter the salinity of water trapped in the surface sediment during subaerial exposure. The only protection from these adverse environmental conditions is to burrow into the sediment. However, in many tidal flats, particularly in muddy sediments, conditions become anaerobic within 1 or 2 cm of the surface. Thus the tidal flat environment is one of extremes for foraminiferids.

North Sea coast Haake (1962) studied tidal flats between Langeoog and the mainland, and Richter (1964a, b, 1965, 1967) those of Jade Bay. Haake recorded current velocities of 0.15 average and 0.33 m/s. maximum. The tidal range is about 2.4 m. The tidal flats are sands and muddy sands.

Name used here	Haake's name
Protelphidium anglicum (Murray)	*Nonion depressulus* (Walker and Jacob)
Nonion depressulus (Walker and Jacob) emend Murray, 1965b.	*Nonion umbilicatulum* (Walker and Jacob)
Elphidium articulatum (d'Orbigny)	*Elphidium excavatum* (Terquem)
Elphidium gunteri (Cole)	*Elphidium gunteri* (Cole)
Elphidium excavatum (Terquem)	*Elphidium selseyense* (Heron-Allen and Earland)
Quinqueloculina dimidiata (Terquem)	*Quinqueloculina seminulum* (Linné)
Ammonia beccarii (Linné)	*Streblus batavus* (Hofker)

The dominant species is *Protelphidium anglicum* (Murray). Although it is widely distributed and not obviously controlled by facies variation, Haake recorded a correlation between abundance and growths of the diatom *Gyrosigma balthica*.

The second most abundant species, *Elphidium excavatum* (Terquem), occurs everywhere but is particularly common on sandflats. It shows a reciprocal relationship with *Protelphidium anglicum* (Murray). The other species are confined to particular substrates. *Nonion depressulus* (Walker and Jacob) lives on mixed muddy sand and sandflats. *Elphidium articulatum* (d'Orbigny) is most common on the muddy sand of the central platform.

Jade Bay is almost marine in its outer part (salinity 32 per mille) and hyposaline in its inner part (salinity 24 per mille). The same species occur as in the Langeoog region. Richter (1964a) summarized the distributions:

High water mark: maximum standing crop. *Protelphidium anglicum* Murray—*Elphidium articulatum* (d'Orbigny) assemblage.

Open tidal flats: small standing crop. Mixed fauna showing affinities with that at low water.

Low water mark and channels: standing crop intermediate between tidal flats and high water mark. *Protelphidium anglicum* Murray—*Elphidium excavatum* (Terquem) assemblage.

Tidal flats slope, creeks: small standing crop. Assemblage as at low water.

It seems probable that *Elphidium articulatum* (d'Orbigny) favours areas of weak currents on sheltered fine sand or muddy tidal flats. *Elphidium excavatum* (Terquem) favours areas of stronger currents and sediment movement. This species lives at a depth of 0.5 to 6 cm in the sediment.

The highest standing crops are found at high water mark regardless of sediment type. Richter (1964b) noted that tidal currents are weakest at low and high water marks. Waves are weakest at high water level because of the large extent of shallow water to be traversed. Also they affect the high water area for less time. Thus the high water level is an area of low energy, and this probably explains the high standing crops.

Possible large-scale transport of living foraminiferids is suggested by the presence of 60 to 240 individuals per 10 cm³ of sediment in transit on the flood tide. *Protelphidium anglicum* Murray and *Elphidium articulatum* (d'Orbigny) are preferentially transported by this method. These two species were also observed in culture to adhere to the underside of the water surface with their pseudopodia. Under calm conditions they could be transported landwards by the rising tide (Richter, 1965).

Transport of living foraminiferids took place during a hurricane in February 1962. This caused the deposition of a 3 to 5 cm sediment layer on reclaimed land. At Voslapp, 71 individuals were present in 40 cm³ of sediment (Richter, 1965). Ten days after the flood, specimens that had been buried deeper than 4 cm had managed to migrate up to the surface 4 cm layer.

Ice can also transport living foraminiferids by the repeated stranding and floating off of ice flakes. The included sediments contain foraminiferids with values of up to 600 per 10 cm³. Richter estimates that at high water near Hooksiel in the winter of 1962–63, 3.5 to 4 million living foraminiferids were present in the 2 m-thick ice accumulation.

To summarize, the hyposaline North Sea tidal flats are characterized by only a few species. Although no quantitative data are available, it seems probable that the diversity values are not greater than $\alpha = 3$. Miliolina are rare or absent, and the Textulariina are not common. The majority of assemblages are 100 per cent Rotaliina.

HIGH ENERGY

Normally, high-energy coasts face on to extensive open seas. Two principal types can be recognized: sandy beaches and rocky coastlines.

A sandy beach is subject to both the pounding of waves and exposure to the atmosphere. It is an adverse environment for epibionts and the fauna normally consists exclusively of endobionts. By contrast, the rocky shoreline offers small scope for endobionts (except a few borers) but in temperate regions seaweeds flourish and many epibionts attach themselves to the rock. Thus, whereas the sandy beach looks barren except for worm casts, the rocky coast looks fertile because of its attached flora and fauna. In the shelter of the latter are many habitats suitable for foraminiferids and other micro-organisms.

Quantitative information available on the foraminiferid assemblages of high-energy coasts is small.

Myers (1943a) made some general remarks concerning the occurrence of foraminiferids on beaches. He pointed out that the intertidal forms could feed only on the microflora of the beach whereas subtidal forms had the opportunity to feed on planktonic organisms as well. *Elphidium crispum* (Linné) was found to grow 40 per cent faster and to reach a diameter 60 per cent larger in the subtidal as compared with intertidal pools in the Plymouth area.

SANDY BEACHES

California   An extensive study of intertidal foraminiferids was undertaken by Reiter (1959). From seven localities around Santa Monica Bay, samples were collected weekly during the period 22 September 1956 to 18 April 1957. These showed seasonal changes in the foraminiferid assemblages. From September to November, the number of species and number of individuals were high. A gradual decrease took place from December to March and an increase started in April. The decrease in abundance is due partly to storms and partly to temperature changes limiting reproduction.

Reiter carried out size analysis of the sediment. He found that an inverse relationship existed between the median diameter of the sand and the number of foraminiferids. With an increase in median diameter to 0.500 mm, there was a decrease in the number of foraminiferids, and *vice versa*. He attributed this to the winnowing of foraminiferids from coarser sands due to their settling velocity being too slow for the conditions of deposition. The winnowed tests are transported seawards to accumulate in the finer sands.

Of the 129 species recorded from the area, only 17 were found living. *Rotorbinella lomaensis* (Bandy), *R. turbinata* (Cushman and Valentine), *Discorbis monicana* Zalesny and *Trochammina pacifica* Cushman live clinging to the substrate and have flat tests. *Cibicides fletcheri* Galloway and Wissler, *Dyocibicides biserialis* Cushman and Valentine and *Planorbulina mediterranensis* d'Orbigny are permanently attached forms. Free living species include *Elphidium spinatum* Cushman and Valentine, *E. translucens* Natland, *Buliminella elegantissima* (d'Orbigny), *Trifarina semitrigona* (Galloway and Wissler) (=*Angulogerina semitrigona* of Reiter), *Quinqueloculina jugosa* Cushman, *Q. lamarckiana* d'Orbigny, *Scutuloris durandi* (Cushman and Wickenden) and *S. redondoensis* Reiter.

Commonly only two species form at least 80 per cent of the living assemblage. The samples were of more or less uniform size although of unknown area. The number of living forms per sample varied from 0 to 1858, but except at two protected stations, the range was 0 to 37. The number of dead individuals showed parallel ranges of 0 to 2280 and 0 to 145. A number of foraminiferids reworked from adjacent outcrops of Miocene and Pliocene rocks were present in the samples. They constitute over 80 per cent of the species but they form less than 25 per cent of the total assemblage.

Cooper (1961) took samples from San Diego, California, northwards to the Columbia River, Oregon, at approximately fifty-mile intervals. Samples from beaches consisted mainly of sand; those from tidepools consisted of algae and, where present, of sediment too. All the tidepools were marine, and regularly flooded at high tide.

More than 120 species were identified, and 64 of these had living representatives. Twenty-two were fossil species reworked from Miocene and Pliocene sediments. When counting the foraminiferids, Cooper did not separate living and dead individuals of the same species as he carried out a total population count. This means that his data are not entirely comparable with those of other authors. However, as he is the sole source of quantitative data, his results are included here with additional interpretation.

The beach faunas can be separated into three provincial faunas plus a group of cosmopolitan species. The southern fauna is dominated by *Cibicides fletcheri* Galloway and Wissler, *Rotorbinella lomaensis* (Bandy) and *R. turbinata* (Cushman and Valentine). The central part, from Gaviota to the

Columbia River, is dominated by *Buccella tenerrima* (Bandy), *Cassidulina limbata* Cushman and Hughes, *Elphidiella hannai* (Cushman and Grant) and *Eponides columbiensis* (Cushman). North of the Columbia River the main forms are *Discorbis ornatissimus* Cushman and *Elphidium microgranulosum* (Thalmann).

Cosmopolitan species include *Trochammina kellettae* Thalmann, *Discorbis monicana* Zalesny, *Elphidium translucens* Natland and, less commonly, *Miliolinella circularis* (Bornemann) and *Quinqueloculina ackneriana* d'Orbigny.

The tidepool faunas also show a latitudinal zonation. In the south the dominant species are *Discorbis campanulata* Galloway and Wissler, *Rotorbinella lomaensis* (Bandy), *Haplophragmoides canariensis* (d'Orbigny) and *Miliolinella circularis* (Bornemann). From Laguna to Gaviota *Glabratella pyramidalis* (Heron–Allen and Earland) is the dominant form, while from San Francisco to Point Arena the main species are *Alveolophragmium columbiense* (Cushman), *Buccella tenerrima* (Bandy), *Discorbis opercularis* (d'Orbigny), *Elphidiella hannai* (Cushman and Grant) and *Elphidium translucens* Natland. The northernmost part is dominated by *Discorbis ornatissimus* Cushman and *Elphidium microgranulosum* (Thalmann). The cosmopolitan group includes *Trochammina kellettae* Thalmann, *Cibicides fletcheri* Galloway and Wissler, *Discorbis monicana* Zalesny, *Quinqueloculina ackneriana* d'Orbigny and *Q. lamarckiana* d'Orbigny.

The Fisher diversity indices are $\alpha = 3$ to $\alpha = 7$ for beach samples and $\alpha = 4$ to $\alpha = 13$ for tidepools. Beach assemblages lack Miliolina and are dominated by Rotaliina. Tidepools have up to 30 per cent Miliolina (Figure 18).

New England   The several beach environments of Martha's Vineyard Island were studied by Todd and Low (1961). Open sea beaches facing the Atlantic Ocean (temperature 19–23 °C, salinity 31.1–32.1 per mille at the time of sampling) are dominated by *Quinqueloculina seminulum* (Linné), *Q. auberiana* d'Orbigny, *Poroeponides lateralis* (Terquem), *Ammonia beccarii* (Linné) and *Elphidium clavatum* Cushman. The attached species, *Rosalina columbiensis* (Cushman) is common on seaweeds growing on boulders in tidepools.

Beaches facing on to Vineyard Sound are dominated by *Rosalina columbiensis* (Cushman), *Elphidium margaritaceum* Cushman, *Miliolinella subrotunda* (Montagu) (=*Quinqueloculina subrotunda* of Todd and Low) and *Quinqueloculina seminulum* (Linné). These beaches are less exposed to wave attack than those facing the Atlantic (temperature 19–20 °C, salinity 31.1–31.6 ° per mille at the time of sampling). They are composed of pebbles and boulders with algal coverings that provide a sheltered habitat for the living foraminiferids.

Other sheltered pebble-cobble beaches facing on to Nantucket Sound have a similar assemblage with the addition of *Ammonia beccarii* (Linné).

Gulf of Mexico   Segura (1963) took samples from nineteen traverses across the beach 'Playa Washington', south-east of Matamoros, Mexico. On each traverse he sampled the intertidal zone, 1 m, 5 m and 15 m depths. The sediments are coarse to fine sands with some muddy material. Living foramini-

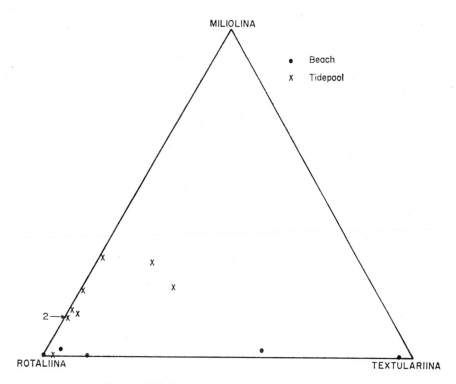

Figure 18 Triangular plot for beach and tidepool samples.

ferids are rare in the intertidal zone and at 1 m depth. They increase to 1–8 per 10 cm² at 5 m to 5–18 per 10 cm² at 15 m. The principal forms are *Ammonia beccarii* (Linné) and varieties, *Elphidium gunteri* Cole, *Quinqueloculina seminulum* (Linné) and *Triloculinella obliquinoda* Riccio. The total (living plus dead) assemblages range from 41 to 6480 per 10 cm³. Segura notes that destruction of tests is common along the strand line and that at a depth of 1 m forms greater than 149 μm in diameter predominate. With increase in water depth, there is an increase in individuals smaller than 149 μm in diameter.

South America: Atlantic coast Boltovskoy (1963a) recorded twenty-eight living species from Puerto Deseado, Patagonia. The dominant form is *Elphidium macellum* (Fichtel and Moll). Seasonal changes in its abundance (Boltovskoy, 1964) are described on pages 204–5.

A study of littoral occurrences (0–15 m depth) along the coasts of Argentina, Uruguay and southern Brazil (Boltovskoy, 1970) shows it to be a single zoogeographic unit. The characteristic species is *Buccella peruviana* (d'Orbigny). This is also the dominant form on the Argentine shelf. Within the major unit there are subunits characterized by *Elphidium discoidale* (d'Orbigny), (North Patagonian, 32° to 41 °S), *Elphidium macellum* (Fichtel and Moll) (South Patagonian, 41° to 52 °S) and *E. macellum, Cassidulina crassa* d'Orbigny, *Trifarina angulosa* (Williamson), *Uvigerina bifurcata* d'Orbigny, *Cassidulinoides parkerianus* (Brady) and *Ehrenbergina pupa* (d'Orbigny) (Malvinian, 52 °S to Tierra del Fuego).

South America: Pacific coast   Four samples from the coast of Chile were dominated by *Glabratella pilleolus* (d'Orbigny) and *Eponides meridionalis* Cushman and Kellett. Locally abundant forms include *Quinqueloculina seminulum* (Linné), *Buliminella elegantissima* (d'Orbigny) and *Buccella peruviana* (d'Orbigny). Boltovskoy and Theyer (1970) noted the contrast between the faunas of the Pacific and Atlantic coasts of South America and particularly the rarity of *Elphidium* on the Pacific coast.

At Ushuaia, Tierra del Fuego, only *Eponides meridionalis* Cushman and Kellett was found alive out of a total of sixty-nine species (Lena, 1966).

Israel   Perath (1966) made a reconnaissance study of the beach sands at Achziv. The tidal range was 40–50 cm during the time of collection (early summer, 1964). The beach is partly protected on the seaward side by a beach-rock barrier which is exposed at low tide.

Shell sands from the 'midlittoral', i.e. the water line of medium high tides, include 15 per cent foraminiferid tests comprising thirty species. Particularly common forms are *Peneroplis planatus* (Fichtel and Moll), *Marginopora* cf. *vertebralis* Quoy and Gaimard and miliolids. However, of these only a few *P. planatus* contained protoplasm and as this was patchily distributed Perath concluded that it was being decomposed. All the shell debris, including the foraminiferids, shows a high polish.

Samples from solution hollows at the midlittoral are much the same as from the shell sands except that the tests are more damaged.

In samples taken at a depth of 1–2 m in a small sand inlet protected by a rocky barrier, foraminiferids form up to 25 per cent of the sand. They are well preserved and less polished. The protoplasm of the 'living' *Peneroplis planatus* filled the chambers.

Active living *Peneroplis planatus* were seen on a calcareous sponge encrusting a small stone in a submerged solution hollow.

In this well agitated and presumably high-energy beach zone, foraminiferids are an important component of the sand, and *P. planatus* forms up to 10 per cent of the grains, although few living individuals have been observed. Perath concludes that the foraminiferids live offshore, and that on death, decomposing individuals with gas-filled tests become weightless and are readily transported to the shore.

Fiji   The only record from this area describes the occurrence of *Marginopora vertebralis* Quoy and Gaimard on a muddy sandflat adjacent to a mangrove swamp (Smith, 1968). Salinities are believed to be variable due to heavy rainstorms.

This species extends from the seaward edge of the mangrove swamp (where it undergoes some hours of subaerial exposure) down to a depth of 1 m below low water. All specimens are attached to a substrate and immature specimens are clumped. Smith also reports the occurrence of this species in the intertidal zone of Borneo.

New Zealand   Hedley, Hurdle and Burdett (1967) described sixty-three species from the intertidal zone. The material was collected from the calcareous alga *Corallina officinalis* over a 1000-mile latitudinal range. The main feature of ecological interest is that the authors concluded that there was no significant faunal change over this latitudinal range.

Summary of living assemblages   Most high-energy sandy beaches face on to open sea with salinities close to normal. This enables miliolaceans to in-inhabit them. Typical genera are *Quinqueloculina* and *Miliolinella*. The principal species are commonly of *Elphidium*, *Ammonia*, *Rosalina* and *Cibicides*. Beyond this it is impossible to define beaches by a set of species. Most beaches seem to be populated by species from the nearby sublittoral areas and these faunas vary according to the geographical position. Diversities are generally low, α less than 7, although in tidepools they rise to a maximum of $\alpha = 13$.

Postmortem changes on sandy beaches   The most noticeable postmortem change is the destruction of tests by abrasion. This leads to the loss of small species and causes the larger, more resistant forms to be damaged. Diversity is very much reduced and foraminiferids are sometimes completely absent from exposed beach sands. Other influences which operate are the introduction of allochthonous foraminiferids either from offshore or from older sediments exposed along the shore.

ROCKY COASTS

Two rockpools near Aberystwyth, Wales, were sampled by Atkinson (1969). The high-level pool had a temperature of 5.5 to 8.5 °C and salinity 30.2 to 34.5 per mille at the time of sampling. For the mid-level pool, the values were 6.0 to 8.5 °C and 31.0 to 34.6 per mille. Both pools are flooded by the tide twice daily. Many different types of algae were sampled, including seaweeds and coralline algae. Further samples were taken of subtidal *Laminaria*.

The principal living species are *Elphidium crispum* (Linné) *Elphidium macellum* (Fichtel and Moll), *Cibicides lobatulus* (Walker and Jacob), *Planorbulina mediterranensis* d'Orbigny, *Massilina secans* (d'Orbigny) and *Rosalina globularis* d'Orbigny. Many other species are present in low abundance.

The total number of individuals on different types of alga is reasonably constant except for on *Enteromorpha clathrata*, *Laurencia pinnatifida* and *Laminaria* spp., all of which support much larger numbers. *Fucus serratus*, *Pelvetia canaliculata* and *Polysiphonia* sp. supported few living foraminiferids.

Among the dominant species *Massilina secans* (d'Orbigny) and *Cibicides lobatulus* (Walker and Jacob) are mainly present on sublittoral seaweeds. Atkinson noted the general absence of living foraminiferids in the adjacent shelf sea, and concluded that the rockpool faunas are indigenous and are not transported in from deeper water.

On death, foraminiferids from rockpools are either destroyed by abrasion or transported offshore into a region of sediment deposition.

3. Estuaries and Lagoons

Excellent reviews of lagoons and estuaries were presented by Emery and Stevenson (1957) and by Hedgpeth (1957b). The more recent data are summarized in Lauff (1967) and Phleger (1969).

No clear separation can be made between lagoonal and estuarine environments in general, but for convenience the following definitions are used:

Estuary—the area of meeting of a river and the sea, an area having rapid environmental changes over short lateral (and sometimes vertical) distances.

Lagoon—a shallow body of water showing some degree of isolation from the open sea.

Lagoons can be subdivided according to their salinity conditions into hyposaline (brackish), normal marine, and hypersaline. Estuaries, of course, are hyposaline, and they include examples that are also hyposaline lagoons as defined above.

More studies of lagoonal and estuarine living foraminiferids have been undertaken than for any other environment, so much of the data has been tabulated here for brevity. A review was published by Murray (1968b).

ESTUARIES

Data are available for the following:

> Estuaries around Hudson Bay and James Bay, Canada
> Miramichi River, Canada
> James River, Chesapeake Bay, U.S.A.
> Rappahannock estuary, Chesapeake Bay, U.S.A.
> Lagoa dos Patos, Rio Grande do Sul, Brazil
> Rio Quequén Grande, Argentina
> Dollart and Ems estuaries, Netherlands
> Dovey estuary, Wales
> Tamar estuary, England
> Christchurch Harbour, England
> Pollution

The general characteristics of these estuaries are summarized in Table 6. Schafer (1969) recorded eight living species in estuaries around Hudson Bay and James Bay:

Astrononion gallowayi Loeblich and Tappan
Islandiella islandica (Nørvang)
Islandiella norcrossi (Cushman)
Elphidium clavatum Cushman

Elphidium incertum (Williamson)
Elphidium orbiculare (Brady)
Elphidium bartletti Cushman
Guttulina dawsoni Cushman and Ozawa

Table 6 Summary of data on estuaries

Location and author	Miramichi River, Canada Bartlett (1966) Tapley (969)		James River Nichols and Norton (1969)	
Sediment				
Sampler	—		—	
Type	Sand, sandy silt, silty sand, silt, silty mud		Silty sands and clays	
Surface layer	pH 6.2–7.0		—	
Disturbance	Reworking		—	
Salinity (per mille)	Upper estuary	Lower estuary	Upper	Lower
Diurnal variation	up to 5	—	—	—
Summer	} 7.86–20.89	23.11–27.44	} 0.5–14	>14
Autumn				
Winter	—	—		
Spring	—	—		
Temperature (°C)				
Diurnal variation	June, July, nearshore areas15		—	—
Summer	20 maximum	20 maximum	28	
Autumn	—	—	—	
Winter	—	—	0	
Spring	<6	<6	—	
Depth (m)	0–15, average 4		6–20	
Tides				
Number	Semidiurnal/diurnal		Tidal	
Amplitude (m)	—		—	
Foraminiferida			Upper river	Lower river
Main species	Miliammina fusca (*Thecamoebina)	Elphidium clavatum	Ammobaculites crassus	Elphidium clavatum
		E. margaritaceum E. orbiculare Eggerella advena important at bay mouth		
Fisher α indices	<1–1.5		—	—
Standing crop per 10 cm²	av. 13	av. 17	5–1500	av. 226

Location and author	Rappahannock Estuary Ellison and Nichols (1970)	Lagoa Dos Patos, Brazil See text
Sediment		
Sampler	—	—
Type	Silty clay	Fine sands, silts, mud
Surface layer	—	? Oxidized
Disturbance	—	Small

Table 6—contd.

Location and author	Rappahannock Estuary Ellison and Nichols (*1970*)		Lagoa Dos Patos, Brazil See text
Salinity (per mille)			
Diurnal variation	—		—
Summer			0.5–26.9
Autumn	0 at head to 16.5 at		6.4–29.7
Winter	mouth		5.8–25.7
Spring			2.3–32.7
Temperature (°C)			
Diurnal variation	—		—
Summer	28		23–27
Autumn	—		22–24
Winter	4		14–20
Spring	—		18–24
Depth (m)	5–25		0–8
Tides			
Number	—		None
Amplitude (m)	0.3–0.8		—
Foraminiferida	Upper estuary	Lower estuary	
Main species	*Ammobaculites crassus*	*Elphidium clavatum*	*Miliammina fusca* *Elphidium gunteri* *Ammonia beccarii* agglutinated forms
Fisher α indices	—	—	<1
Standing crop per 10 cm²	av. 21		<20–>100

Location and author	Rio Quequén Grande Wright (*1968*)	Dollart–Ems Ván Voorthuysen (*1960*)	Dovey Estuary Haynes and Dobson (*1969*)
Sediment			
Sampler	—	—	—
Type	Sands and clayey sands	Silt	Sand
Surface layer	—	—	—
Disturbance	reworked Holocene	—	reworked by currents
Salinity (per mille)			
Diurnal variation			
Summer			
Autumn	0–12 Upper	11–29	3–33
Winter	32–35 Lower		
Spring			

Table 6—contd.

Location and author	Rio Quequén Grande Wright (1968)	Dollart–Ems Van Voorthuysen (1960)	Dovey Estuary Haynes and Dobson (1969)
Temperature (°C) Diurnal variation Summer Autumn Winter Spring	} 7–25	} 18–24	} 1–17
Depth (m)	1–3	0–10	0–8.5
Tides Number Amplitude (m)	Strong —	Semidiurnal 2.3–3	— 2.4–4.8
Foraminiferida Main species	Miliammina fusca Quinqueloculina millettii	Ammonia beccarii Protelphidium anglicum Elphidium gunteri Elphidium excavatum Elphidium articulatum	rare
Fisher α indices Standing crop per 10 cm²	— 0–20	— —	— —

Location and author	Tamar Estuary, Plymouth, England Murray (1965c)	Christchurch Harbour, England Murray (1968a)	
Sediment Sampler Type Surface layer Disturbance	Mud, muddy sand, sand Oxidized Locally reworked	Silty sand, sand Oxidized Small	
Salinity (per mille) Diurnal variation Summer Autumn Winter Spring	Grab up to 10 33–35 — 21–31 26–32	Dredge — 14–34 27–35 <1 2–29	
Temperature (°C) Diurnal variation Summer Autumn Winter Spring	up to 5 12.9–15.3 — 7.9–8.0 9.4–14.4	— 17.6–19.8 14.9–19.0 4.8–9.7 11.0–13.0	
Depth (m)	0–38	0–5, average 2	

Table 6—contd.

Location and author	Tamar Estuary, Plymouth, England Murray (1965c)	Christchurch Harbour, England Murray (1968a)	
Tides			
Number	Semidiurnal	4 high water and 4 low water daily	
Amplitude (m)	4.73 springs, 2.34 n	1.5 springs, 0.6 neaps	
		Lower estuary	Upper estuary
Foraminiferida Main species	*Protelphidium anglicum* *Ammonia beccarii* *Elphidium* spp.	*Protelphidium anglicum* *Elphidium* sp. plus in autumn *Ammonia beccarii* *Elphidium articulatum* *Reophax moniliformis* plus in summer *Miliammina fusca* *E. oceanensis* *E. articulatum* *A. beccarii* *R. moniliformis*	*Protelphidium anglicum* *Elphidium* sp. plus in autumn *Miliammina fusca* *Elphidium oceanensis* plus in summer *Miliammina fusca* *Ammonia beccarii*
Fisher α indices	3–5	1–5	<1–3
Standing crop per 10 cm²	not known	not known	

All were rare. No details were given of temperature, salinity or substrate. Schafer commented on the absence of *Miliammina fusca* (Brady) and the northern aspect of the *Elphidium* species.

Miramichi River drains into the Gulf of St Lawrence. It has a considerable salinity variation. The upper estuary is dominated by *Miliammina fusca* (Brady). In the lower estuary *Elphidium* dominates, notably *E. clavatum* Cushman, *E. margaritaceum* Cushman and *E. orbiculare* (Brady). At the estuary entrance *Eggerella advena* (Cushman) is important. Diversity is very low (α less than $1\frac{1}{4}$, Figure 19.) The standing crop increases from an average of 3.5 per 10 cm² in the river to 13 in the estuary and 17 in the mouth.

The James River estuary opens into Chesapeake Bay. During the time it was sampled (May–August 1967) the salinity was 12–16 per mille, with occasional extremes 9.8–17.9 per mille, in the central part. Temperature was 13.3–27.8 °C and oxygen 4.5–8.6 cm³/l. The upper estuary is dominated by *Ammobaculites crassus* Warren and a variant, with subsidiary *Miliammina fusca* (Brady), and *Ammobaculites* spp. (Nichols and Norton, 1969). The lower estuary has salinities greater than 14 per mille. *Elphidium clavatum* Cushman is the dominant species. It is accompanied by subsidiary *Ammobaculites crassus* Warren, *Trochammina squamata* Parker and Jones, and *Tiphotrocha comprimata* (Cushman and Brönnimann). No tables of data are available to determine α values or position on the triangular plot (no miliolaceans are present). Upstream from the *Ammobaculites crassus* assemblage there are Thecamoebina.

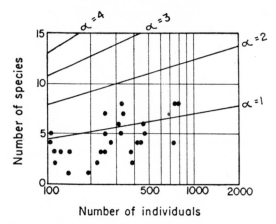

Figure 19 Diversity plot for Miramichi River (data from Bartlett (1966) and Tapley (1969).

Another estuary opening into Chesapeake Bay, Rappahannock estuary, has similar features (Nichols and Ellison, 1967; Ellison and Nichols, 1970). The salinity varies from zero near the head to an average of 16.5 per mille at the mouth. Daily fluctuations of 5–13 per mille are experienced in the gradient zone between the lower and upper estuary. *Ammobaculites crassus* Warren dominates the upper and *Elphidium clavatum* Cushman the lower estuary, with subsidiary species *Miliammina fusca* (Brady) and *Ammonia beccarii* (Linné). The standing crop is low, on average 21 per 10 cm², but reaches 500 per 10 cm² in parts of the upper estuary. Where eel-grass grows, *Ammonia beccarii* is very abundant. Milioaceans are absent.

Lagoa dos Patos, Rio Grande do Sul, Brazil, is a large non-tidal lagoonal estuary. The salinity range is extreme (0–28 per mille) but spread over a considerable lateral distance. Closs (1963) recognized five salinity zones from the southern end of the lagoon (zona prélímnica, salinity 0–19 per mille) with *Miliammina fusca* (Brady) and Thecamoebina, through the estuary proper to the entrance to the sea (zona pré-marinha, salinity 0–28 per mille) with a fairly marine fauna. Most of the estuary is dominated by *Miliammina fusca* (Brady), *Elphidium* spp. (particularly *E. gunteri* Cole), *Ammotium cassis* (Parker) and *Ammonia beccarii* (Linné). The standing crop varies from less than 20 to greater than 100 per 10 cm². Similar results were recorded by Closs and Madeiros (1965).

Forti and Roettger (1967) took samples twice a week from this estuary. Salinity ranged from 0.46 to 26.91 per mille in the summer to 2.34 to 32.76 per mille in the spring. Temperature reached a winter minimum of 14 °C and a summer maximum of 27 °C. Living foraminiferids were most abundant in the winter, and this is believed to be the period of growth of calcareous species. The agglutinated forms reproduce during the winter. Abnormal tests were noticed in *Elphidium gunteri* Cole and *E. incertum* (Williamson) in the winter and in *Textularia earlandi* Parker and *Trilocularena patensis* Closs during the spring. A similar study was carried out at two stations by Closs and Madeira (1968). They concluded that the time of reproduction of individual species was controlled more by the quantity of food available than by the salinity and temperature conditions.

In general the results presented for the Lagoa dos Patos cannot be plotted on a triangular diagram or for α. The three samples, with more than 100 individuals, studied by Closs and Madeiras (1968), are 100 per cent Rotaliina (2), 100 per cent Textulariina (1) and have α values below 1.

In the estuary of the Rio Quequén Grande, Argentina, only the living miliolids were studied (Wright, 1968). The estuary is shallow (1–3 m depth) and subject to powerful tides. Salinity varies from 0–12 per mille at the head to 32–35 per mille at the mouth. In the central part the range is 4–30 per mille and the diurnal change may be as high as 23 per mille. The annual temperature range is 7–25 °C.

The standing crop of miliolids is in the range 10–20 per 10 cm² in the mouth region where salinities are 32–35 per mille. In the central parts of the estuary, values of 0–5 per 10 cm² are found in the lower-salinity waters. These values include *Miliammina fusca* (Brady) which is here regarded as a textulariian. This form is the principal species in the central part of the estuary.

Samples collected seasonally from the same place at the mouth of the estuary reveal maximum abundance in the summer (up to 50 per 10 cm²), moderate abundance in the autumn (up to 28 per 10 cm²), and sparse occurrence during winter and spring. In the latter two seasons salinity is generally lower, but although temperature is lowest in winter it rises in the spring.

Van Voorthuysen (1960) took Van Veen grab samples from the intertidal and subtidal zones of the Dollart and Ems estuaries in the Netherlands. These are areas of fairly big tides and large ranges of temperature and salinity. The samples were split into four fractions: greater than 1 mm containing no foraminiferids, 0.3–1 mm with few foraminiferids, 0.15–0.3 mm with abundant foraminiferids, and 0.05–0.15 mm with juveniles, elongate tests such as *Bolivina* and reworked Cretaceous species.

Five species are dominant throughout the area:

Name used here	*Van Voorthuysen's name*
Ammonia beccarii (Linné)	*Streblus batavus* Hofker
Protelphidium anglicum Murray	*Nonion depressulus* (Walker and Jacob)
Elphidium gunteri Cole	*Elphidium gunteri* Cole
Elphidium excavatum (Terquem)	*Elphidium selseyensis* (Heron–Allen and Earland)
Elphidium articulatum (d'Orbigny)	*Elphidium excavatum* (Terquem)

At any one sample point, only one or two of these occur in abundance. *Elphidium gunteri* is particularly common in the Dollart estuary. Among the rarer species, there are few Miliolacea or Textulariina. The dead assemblages include some forms swept in from the open North Sea.

The Dovey estuary, Wales, has a moderate tidal range so that the sediments are considerably disturbed and reworked during each tidal cycle.

Haynes and Dobson (1969) recorded rare living individuals on the sandy substrate. Where the channel passes by a rocky shore, seaweeds and mussels provide a habitat for a few *Quinqueloculina seminulum* (Linné) and *Miliolinella subrotunda* (Montagu).

The Tamar estuary, Plymouth, England, is deep (0–38 m) and the portion here considered is the lower estuary (Murray, 1965c). There is a seasonal variation in the salinity values, and the diurnal variation at any one point reaches up to 10 per mille. The tidal amplitude is moderate. The dominant species are *Protelphidium anglicum* Murray, *Ammonia beccarii* (Linné) and *Elphidium* spp. The standing crop is thought to be small. Two samples gave α values of 3 and 5.

Christchurch Harbour, England, has been studied in greater detail than any other estuary (Murray, 1968a). This small estuary has an average depth of 2 m. There is an extreme seasonal salinity cycle, and during some wet years the salinity remains too low for foraminiferids to continue to live there. The dominant species change from one season to the next (Table 6). The important forms are *Protelphidium anglicum* Murray, *Elphidium* sp., *E. articulatum* (=*E. excavatum* of Murray, 1968a), *E. oceanensis* (d'Orbigny) and *Miliammina fusca* (Brady).

The seasonal environmental changes are summarized in Table 7. The diversity values are plotted in Figure 20. In the lower estuary diversity is highest in the summer (α = 1 to α = 5), followed by autumn (α = 2 to α = 4) and spring (α = 1 to α = 4) while during the winter it falls to α = 1 to α = 2. These changes parallel the environmental changes. In the

Table 7 Seasonal environmental changes, Christchurch Harbour, England

Upper Estuary

Season	Temp. (°C)	Salinity (per mille)	Oxygen (cm³/l)	pH
Autumn	16.2–18.5	7–32	1.88–4.56	7.8–8.4
Winter	4.5–6.5	0–1	1.97–3.70	8.0–8.4
Spring	10.8–15.4	0–30 (generally <10)	2.65–5.25	7.0–8.3
Summer	17.3–20.0	3–33	2.65–4.45	7.5–8.3

Lower Estuary

Season	Temp. (°C)	Salinity (per mille)	Oxygen (cm³/l)	pH
Autumn	17.1–19.0	20–35	2.91–5.32	7.9–8.4
Winter	4.4–9.9	0–1 (rarely up to 29)	1.51–4.52	7.4–8.4
Spring	10.1–12.3	14–33	2.70–4.79	7.1–8.2
Summer	17.4–19.7	32–35 (rarely 7–35)	2.70–5.03	7.6–8.3

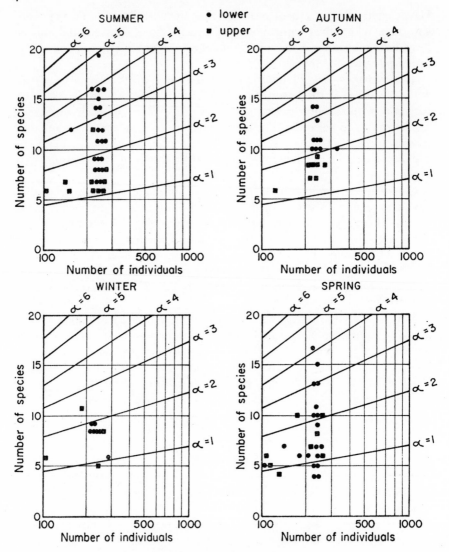

Figure 20 Diversity plot for Christchurch Harbour
(data from Murray, 1968a).

upper estuary diversity is consistently low throughout the year ($\alpha > 1$ to $\alpha = 3$) because the environment is always variable.

Murray (1968a) used a method of comparing the range of environmental parameters with the abundance of the species. The percentages of the species at each station of occurrence are arranged in order of increasing abundance from left to right and plotted as histograms. In the upper part of each seasonal sector, temperature and depth data are plotted (the depth curve in no way indicates a bottom profile). In the lower part of each sector are plotted the chlorinity, pH, dissolved oxygen content, and calcium and magnesium content of the water immediately above the sediment. An example is shown in Figure 21 for *Quinqueloculina dimidiata* Terquem. This

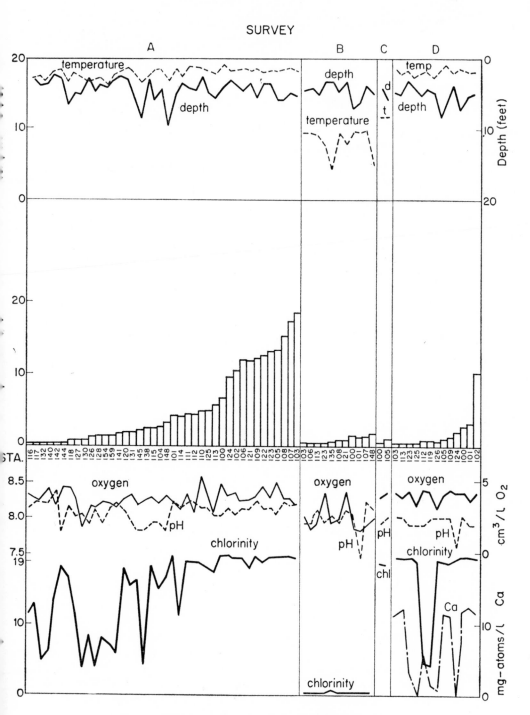

Figure 21 Histograms of abundance of *Quinqueloculina dimidiata* Terquem and environmental parameters.

Table 8 Optimum conditions for the main species, Christchurch
Harbour, England

Species	Optimum salinity (per mille)	Range of salinity tolerated by a population of > 5%	Optimum calcium content of bottom water (mg-atoms/l)	Optimum temperature range (°C)
Eggerella scabra	35	29–35	12	?15–20
Quinqueloculina dimidiata	35	31–35	11–12	17–20
Nonion depressulus	35	0–35	11–12	5–10
Buccella frigida	35	35	probably high	16–19*
Reophax moniliformis	33–35	9–35	not obvious	15–20
Ammonia beccarii	probably >10	0–35	not obvious	?15–20
Elphidium articulatum	29–35	0–35	not obvious	10–15
Miliammina fusca	not obvious	0–35	<2	?10–20
Elphidium oceanensis	35	9–35	not obvious	10–20
Protelphidium anglicum	not obvious	0–35	not obvious	not obvious

* May be anomalous due to unfavourable salinities preventing the full development of this species.

species shows a clear preference for water of near-normal salinity and autumn temperatures (survey A). Using this method together with distribution maps, it is possible to define the optimum and range of tolerance of some of the environmental parameters (see Table 8). It is further possible to recognize four groups of species:

1. Stenohaline marine forms which are transported into the estuary from the open sea.
2. Marine forms able to tolerate small variations in salinity.
3. Moderately euryhaline forms: *Reophax moniliformis* Siddall, *Ammonia beccarii* (Linné) and *Elphidium articulatum* (d'Orbigny) (= *E. excavatum* of Murray, 1968a).
4. Highly tolerant euryhaline and eurythermal forms: *Miliammina fusca* (Brady), *Elphidium oceanensis* (d'Orbigny), *Elphidium* sp. and *Protelphidium anglicum* Murray.

A computer interpretation of the data for the summer season in general confirmed the results described above (Howarth and Murray, 1969).

Summary of estuaries

In general, estuaries can be divided into an upper part subject to the greatest fresh-water influence and a lower part connected with the sea. Where diversity data are available, the α values are very low in the upper estuary (are less than 3) and low in the lower estuary ($\alpha = 1$ to $\alpha = 5$). A common upper estuary dominant species is *Miliammina fusca* (Brady). Often it also occurs in the lower estuary with *Ammonia beccarii* (Linné) and various species of *Elphidium*. There is little information on standing crop size, but it is likely that values are high in estuaries such as Christchurch Harbour during the favourable seasons of the year. A plot of the three suborders (Figure 22) shows a predominance of Rotaliina and to a lesser extent of Textulariina. All the samples with more than 3 per cent Miliolina in Christchurch Harbour are from the lower estuary during the autumn, when salinities were 34–35 per mille. Thus this resembles a normal marine distribution and is not really characteristic of estuaries in general.

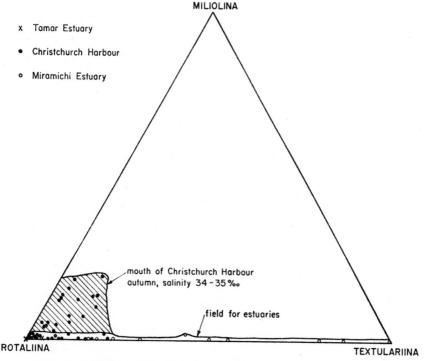

Figure 22 Triangular plot of data for estuaries.

Note on fresh-water foraminiferids

Boltovskoy and Lena (1971) have reviewed the occurrence of foraminiferids in fresh water. In the Rio de la Plata they found *Protelphidium tisburyensis* (Butcher) (=*Nonion tisburyense* of Boltovskoy and Lena) to be common. Species present in small numbers include *Miliammina fusca* (Brady), *Psammosphaera* sp. and *Trochammina* sp. Other species live in fresh-water regions that periodically become hyposaline: *Rotalia beccarii parkinsoniana* (d'Orbigny), *Reophax arcticus* Brady, *Elphidium excavatum* (Terquem), *Haplophragmoides wilberti* Andersen and *Ovammina* sp. Boltovskoy and Lena believe the fresh-water intervals to be greater than the expected life-span of these species. However, they do not prove by observation that the foraminiferids were active or able to reproduce. Murray (1963) showed experimentally that adverse salinities could prevent foraminiferids from feeding. However, they are known to survive for long periods without food. Clearly, further experimental work in this field is required before it can be definitely claimed that foraminiferids live in fresh water.

Pollution

The Restigouche estuary on the east coast of Canada has been studied by Schafer (1970). Samples were taken close to known sources of pollution. Species named as potentially useful indicators of pollution are:

> *For outer estuary*
> *Ammomarginulina fluvialis* (Parker)
> *Elphidium frigidum* Cushman
> *Hemisphaerammina* sp.
> *Pseudopolymorphina novangliae* (Cushman)
> *Reophax fusiformis* Williamson
> *Reophax arctica* Brady
> *Reophax nodulosa* Brady
> *Reophax scottii* Chaster
> *Saccammina atlantica* (Cushman)
> *Trochammina inflata* (Montagu)

> *For inner estuary*
> *Eggerella advena* (Cushman)

Schafer listed changes in the areas of effluent discharge. Most changes were associated with improvement towards the mouth of the estuary. This would be expected under unpolluted conditions so that the recognition of pollution effects is rendered difficult. However, pH values for the water were in the range 4.4 to 7.5 at different outfalls. No foraminiferids were found at pH 4.4 and the most obvious trend in the above list of indicator species is the predominance of agglutinated forms. In this respect these results show similarities with those from polluted areas of the Californian shelf (pages 103–5).

LAGOONS

Hyposaline lagoons

These are discussed in the following order:

North America (Table 9)
  Tracadie Bay, St Margarets and Mahone Bays, Canada, Hadley Harbor, Buzzards Bay, Long Island Sound, Tampa-Sarasota Bay.

South America (Table 9)
   Laguna de Terminos, Mexico; various Brazilian lagoons.
Europe
   Bottsand Lagoon, N. Germany
   Llandanwg Lagoon, Wales
   Arcachon, France
Japan

North America The northernmost lagoon for which data is available is
Tracadie Bay, Prince Edward Island (Bartlett, 1965). This is very shallow.
At the time of sampling salinity showed little variation. The annual tem-
perature changes are extreme (Table 9, Figure 23). St Margarets Bay and
Mahone Bay, Nova Scotia (Bartlett, 1964) have a greater depth range,
consequently conditions are more stable in the deeper parts. (Table 9,
Figure 23). All these bays freeze over during the winter.
   The diversity ranges from below $\alpha = 1$ up to $\alpha = 4$, with the lowest
values on silty substrates in Tracadie Bay and in the marginal parts of St
Margarets and Mahone Bays (Figure 24). On the triangular plot (Figure 25)
all the samples lie along the Rotaliina-Textulariina side. The standing crop
is variable and small, 1–93 per 10 cm². The dominant species are *Elphidium*
spp., *Eggerella advena* (Cushman) and *Miliammina fusca* (Brady) with the
addition of *Ammotium cassis* (Parker) in Tracadie Bay and *Saccammina*
atlantica (Cushman) (=*Proteonina atlantica* of Bartlett) in St Margarets
and Mahone Bays.
   In Hadley Harbor on Cape Cod, Buzas (1968b) recorded six stations
with more than 100 living foraminiferids. The dominant species are *Elphidium*
clavatum Cushman, *E. subarcticum* Cushman and *Ammonia beccarii* (Linné),
and Rotaliina form 97.5–100 per cent. Diversity ranges from $\alpha = 1$ to
$\alpha = 3.5$. Salinities are 31–32 per mille but the annual temperature range is
no doubt large as in Buzzards Bay (see below).
   Buzzards Bay is only slightly hyposaline (salinity 29.5—32.5 per mille),
but there is a large annual temperature variation. Long Island Sound is more
hyposaline (salinity 25–29 per mille) and has a similar temperature range
(Table 9, Figure 23). In Buzzards Bay, Murray (1968b) found that the diversity is different on sand and muddy substrates, with values of $\alpha = 3$ to
$\alpha = 7$ and $\alpha = 1.5$ to $\alpha = 3$ respectively (Figure 24). There is also a corres-
ponding difference in standing crop: average 171 per 10 cm² on sand and
32 on mud. In Long Island Sound (Buzas, 1965), where muddy and silty
sediments prevail throughout, α values are low at $\alpha < 1$ to $\alpha = 2$. Standing
crop values are generally higher but decrease with depth (Table 9). On the
triangular plot, all points lie close to the Rotaliina—Textulariina side
(Figure 25). Miliolaceans are present in Buzzards Bay because the salinity
rises to 32 per mille. They are completely absent from Long Island Sound.
Elphidium clavatum Cushman and *Eggerella advena* (Cushman) are dominant
in both lagoons, together with *Ammonia beccarii* (Linné) and *Fursenkoina*
fusiformis (Williamson) in Buzzards Bay and *Buccella frigida* (Cushman) in
Long Island Sound. Buzas recognized a lengthwise zonation of Long Island
Sound based on variations in abundance of the dominant species.

Table 9. Summary of Hyposaline Lagoon Data

	Tracadie Bay, Prince Edward Island, Canada	St Margarets and Mahone Bay, Nova Scotia, Canada	Buzzards Bay, Mass., U.S.A.	Long Island Sound, U.S.A.	Tampa–Sarasota Bay, Florida	Laguna de Terminos, Mexico
Author	Bartlett (1965)	Bartlett (1964)	Murray (1968b)	Buzas (1965)	Walton (1964b)	Ayala-Castañares (1963)
Sediment						
Type	Silty clays to coarse sand. Well to normally sorted	Silty and fine organic ooze, some glacial boulder clay and coarse sand	Poorly sorted silty sand and muddy silt	Poorly sorted clayey silt and silty sand	Fine to very fine sand with silt in upper bay	Muddy sands with up to about 50% CaCO$_3$
Surface layer	pH 7.2–7.8	pH 6.8–7.2	Oxidized	Oxidized	Oxidized	Oxidized
Disturbance	High winds cause turbulence	Powerful currents at bay entrances	Small Currents < 1 knot	Small	Tidal currents <0.7 knots	
Salinity (per mille)						
Diurnal variation	—	—	—	—	—	—
Summer	At time of sampling 27.84–28.39	May–September 1962–3 back bay 24–27, open bay 26–30, central bay 30–32	Annual variation 29.5–32.5 (Sanders, 1958)	29 maximum	May be < autumn 1952–23.6 at head, 32.1 at mouth, 33–34 Sarasota Bay	—
Autumn						—
Winter				25 minimum	—	—
Spring					—	21–38

			After Sanders (1958) and Moore (1963)			
Temperature (°C)						
Diurnal variation	10					
Summer	26 maximum	depth 3 m, 1–22	20	25 maximum	—	—
Autumn		depth 20 m, 4–10				—
Winter	1 minimum	depth 20–70 m, <4	1	2 minimum	—	—
Spring		frozen December–January			1952. 21–23	26–31
Depth (m)	0–6	0–73 St Margarets 0–50 St Mahone (often <10)	0–30 Range for samples 12–24	0–43	0–10	0–4
Tides						
Number	No data	Diurnal/semidiurnal	Semidiurnal	Semidiurnal	? Semidiurnal	
Amplitude (m)	1–1.3	1.3	1.3		0.5	0.7
Foraminiferida						
Main species	Elphidium incertum, E. orbiculare, Ammotium cassis, Eggerella advena, Miliammina fusca	Elphidium spp., Eggerella advena, Miliammina fusca, Saccammina atlantica	Eggerella advena, Ammonia beccarii, Elphidium clavatum, Fursenkoina fusiformis	Elphidium clavatum, Buccella frigida, Eggerella advena	Upper Bay Ammotium salsum, Ammonia beccarii, Elphidium spp. Lower Bay Elphidium spp., E. matagordanum, Quinqueloculina poeyana, Ammonia beccarii, Triloculina brevidentata	Ammonia beccarii, Elphidium spp.
Fisher α indices	Sand 1.25–3, Silt and clay <1	Back, marginal <1.25, Open, central 2–4	Sand 3–7, Silt and clay 1.25–3	<1–2	1.25–2.5	<1–4
Standing crop	1–72	3–93	0–308, mud av. 32, sand 171	depth <10 m, av. 335, depth 10–20 m, av. 177, depth 20 m, av. 62	6–359, commonly <70	av. <100

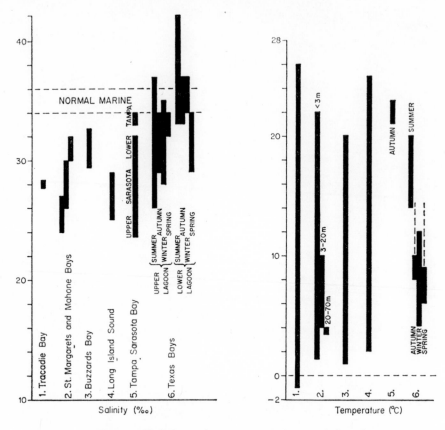

Figure 23 Salinity and temperature characteristics of
some hyposaline lagoons.

Seasonal changes in the abundance of species in Long Island Sound show
a peak for *Eggerella advena* (Cushman) in October, 1961, and for *E. advena*,
Elphidium clavatum Cushman and *Buccella frigida* (Cushman) in June 1962.
At this time the total living assemblage was greatest. Buzas correlates this
with the zooplankton and phytoplankton cycles and with the period of
maximum temperature.

Tampa-Sarasota Bay is on the west side of Florida. It shows a range of
salinities from 23.6 to 34 per mille. Temperature variation is probably not
very great. Diversity values are low: $\alpha = 1.5$ to $\alpha = 2.5$ (Figure 24). On the
triangular plot, four samples lie close to the Rotaliina–Textulariina side with
only a small Miliolina component (Figure 25). One anomalous sample has
64 per cent Miliolina. Its water depth is 0.3 m and the sediment contains
20 per cent shell debris. Its proximity to the bay entrance suggests that the
salinity may be nearly normal.

South America   Laguna de Terminos, Campeche, Mexico (Ayala-Castañares, 1963) is generally shallow except in the tide-swept channels connec-
ting it with the open Gulf of Mexico. Incoming Gulf water of 38–40 per
mille salinity mixes with the somewhat hyposaline water of the lagoon

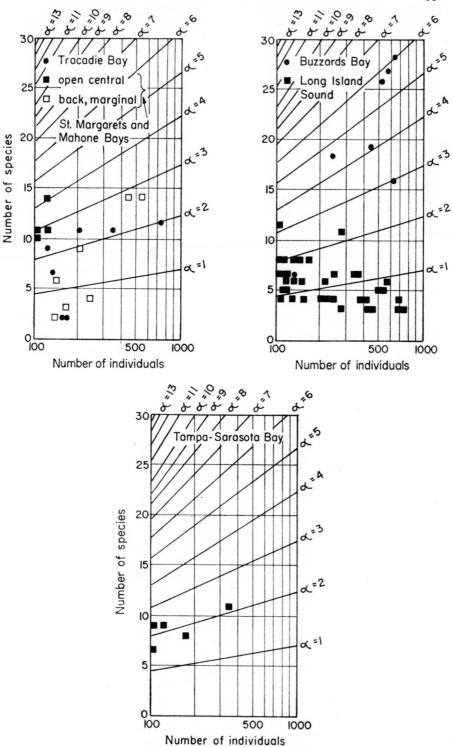

Figure 24 Diversity plots for hyposaline lagoons.

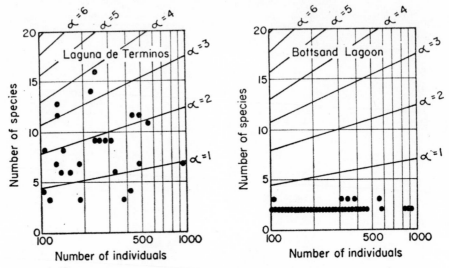

Figure 24 (contd.).

proper (salinity 27–33 per mille except close to the shore where it ranges from 21–27 per mille). The Gulf water also introduces much of the $CaCo_3$ present in the sediment. In the southern part, the water is turbid so that there is no vegetation, but elsewhere there is a flora of *Thalassia testudinum* and *Diplantheria wrightii*.

The standing crop of foraminiferids ranges from 1 to 935 per 10 cm², but most values are less than 100 per 10 cm². The diversity indices are low; $\alpha < 1$ to $\alpha = 4$ (Figure 24). On the triangular plot (Figure 25) all the samples are in the Rotaliina corner. Miliolaceans are rare in the dead as well as in the living assemblage. They clearly do not live on the vegetation here because the salinities are too low. The dominant species are *Ammonia beccarii* (Linné) and *Elphidium* spp.

A series of papers by Closs and Madeira (1962, 1966, 1967), Closs and Madeiros (1967) and Madeira (1969) provide general information on various Brazilian lagoons. Species present include *Ammonia beccarii* (Linné), *Elphidium* spp., *Miliammina fusca* (Brady) and *Ammotium salsum* (Cushman and Brönnimann). The data are not suitable for determining α indices or for use on a triangular plot.

Europe Bottsand lagoon is on the east side of Kiel Bay (Lutze, 1968a). The water is less than 1 m deep. The salinity is 12–18 per mille, although occasional storms cause further lowering by fresh-water land runoff. Temperature ranges from 0 °C in winter (ice present) to 28–30 °C in summer. Because of the small depth the diurnal temperature range in summer is sometimes as high as 10 °C.

The diversity values (Figure 24) are very low. On the triangular plot (Figure 25) all the points lie along the Rotaliina–Textulariina side. Standing crop varies from 1 to 343 per 10 cm², with most values between 100 and 200 per 10 cm². The dominant species are *Elphidium articulatum* (d'Orbigny)

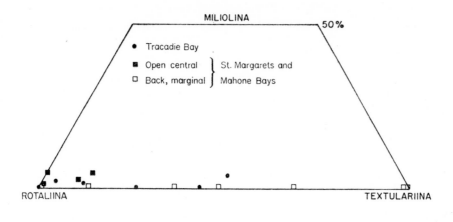

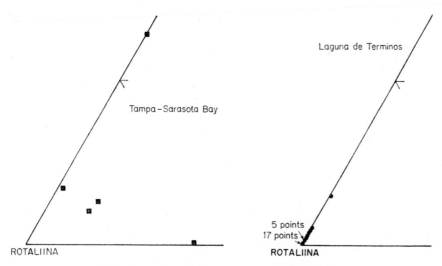

Figure 25 (Part 1). Triangular plots for hyposaline lagoons.

(=*Cribrononion articulatum* of Lutze) and *Miliammina fusca* (Brady). Samples taken during winter, when the lagoon was frozen from top to bottom, contained *E. articulatum* with protoplasm. These may have been alive or just preserved. Both species show a seasonal change in abundance, with minimum standing crop values in late winter (March) and maximum values in November. Similarly, size measurements show that *E. articulatum* reproduces in April and reaches its maximum size in the winter months.

A study of *E. articulatum* (=*E. excavatum* of Haman) in Llandanwg Lagoon, Wales, (Haman, 1969) showed the maximum standing crop to be in May–June. It was concluded that sexual reproduction occurs on a small scale in January during adverse environmental conditions (which cause the death of most juveniles). In April the main asexual reproductive phase starts, and the standing crop reaches its peak in May–June. Another minor reproductive phase occurs in August–September. This lagoon has an annual temperature range of 1–17 °C and a salinity range of 3–30 per mille.

The Bassin d'Arcachon is a large embayment on the Atlantic coast of France. Water temperatures range from a mean value of 7.3 °C in December to 22.4 °C in August. Salinities range from 29 to 35 per mille. Most of the lagoon is normal marine or only slightly hyposaline. The tidal range

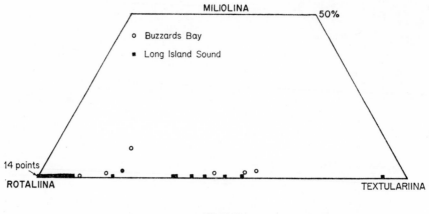

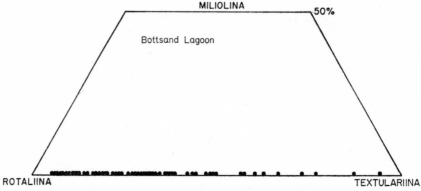

Figure 25 (Part 2).

is 2–4.5 m. At low water, much of the bottom is exposed and water remains only in the deep channels. Le Campion (1970) distinguished a number of subenvironments, but most of his data cannot be used for a triangular plot or to determine α values.

I Subtidal subenvironments

1. Shell-sediments in the channels. These are mainly lag deposits of shell debris winnowed during each tidal cycle. The only living species is *Cibicides lobatulus* (Walker and Jacob).
2. Mussel-beds. These occupy elongate areas near the lagoon entrance. Few species are present, the dominant ones being *Fissurina orbignyana* (Sequenza) (=*Entosolenia orbignyana* of Le Campion), *Planorbulina*

mediterranensis d'Orbigny, *Asterigerinata mamilla* (Williamson) (=*Discorbis mamilla* of Le Campion) and *Textularia earlandi* Parker.

3. *Amphioxus* sands. These occur in the entrance channel. Living foraminiferids are almost completely absent.

4. Calcareous algae. The foraminiferids living in association with *Jania* on rock outcrops are dominated by *Discorbis* cf. *D. globularis* (d'Orbigny) var. *bradyi* Cushman, 84 per cent. Also present are miliolids, 5 per cent. Nearby sediments yield a different assemblage dominated by *Elphidium crispum* (Linné), *E. macellum* (Fichtel and Moll) and *E. poeyanum* (d'Orbigny), *Quinqueloculina* spp. and *Cibicides* spp.

5. Meadows of *Zostera marina*: this plant grows in water less than 3.5 m deep. The living assemblage where the substrate is muddy is dominated by species that cling to the *Zostera*, namely, *Planorbulina mediterranensis* d'Orbigny and *Cibicides lobatulus* (Walker and Jacob). Where the substrate is sand or sandy mud, the assemblage comprises *Massilina secans* (d'Orbigny), *Ammonia beccarii* (Linné) and *Elphidium crispum* (Linné).

6. Infralittoral sands with *Venus* or *Abra*: certain species are found here and not elsewhere in the lagoon. *Crithrionina goesii* Höglund lives attached to dead mollusc shells and to plates of mica. *Nonionella turgida* (Williamson) is abundant during the spring. The main faunal elements are Textulariina (up to 47 per cent, particularly *Eggerella scabra* (Williamson)), and *Planorbulina mediterranensis* (d'Orbigny).

II Intertidal subenvironments

1. Meadows of *Zostera nana*. At low tide areas of this plant become exposed. The sediment is muddy. In regions of low salinity (<24 per mille) *Protelphidium anglicum* Murray (=*Nonion depressulum* of le Campion) and *Elphidium articulatum* (d'Orbigny) (=*E. excavatum* of le Campion) are dominant and *Ammoscalaria pseudospiralis* (Williamson) is common. In regions of higher salinity (>24 per mille) the dominant form is *Eggerella scabra* (Williamson).

2. *Scrobicularia* mudflats: the main species is *Ammoscalaria pseudospiralis* (Williamson), and where salinity is less than 24 per mille the assemblage is the same as in the *Zostera nana* meadows.

Japan Matoba (1970) has described the assemblages of Matsushima Bay. Depths are less than 3 m. Salinity ranges from 25–31 per mille and temperature from 26–29 °C .The bottom is muddy and is colonized by *Zostera*. Standing crop ranges from less than 100 to greater than 300 per 10 cm². However, in only two samples were more than 100 individuals counted to determine the percentage composition of the assemblage. The dominant species are *Trochammina hadai* Uchio, *Trochammina* cf. *T. japonica* Ishiwada, *Ammonia beccarii* (Linné) and *Elphidium subarcticum* Cushman.

Hyposaline to hypersaline lagoons

A general discussion of the marginal marine faunas and floras of the Golfe du Lion has been given by Lévy (1971). He lists the occurrence of foraminiferids in lagoons ranging from fresh through hyposaline and marine to hypersaline. It is not clear whether the observations are based on living or dead foraminiferids. In waters of salinity less than 30 per mille the abundant forms are *Ammonia beccarii* (Linné), *Protelphidium* and '*Cribrononion*' *gunteri*. The miliolids become common at salinities greater than 30 per mille especially on sandy substrates. In hypersaline waters more than 40 per cent salinity) *Ammonia beccarii* (Linné) and *Protelphidium* are again dominant.

Along the Texas coast of the Gulf of Mexico is a series of lagoons in which conditions vary from hypo- to hypersaline according to the season of the year. These are Aransas, Mesquite and San Antonio Bays, described by Phleger (1956) and Phleger and Lankford (1957). A sand-barrier island complex with few openings separates the lagoons from the sea. The main supply of fresh water is the Guadelupe River which flows into San Antonio Bay.

The lagoons are generally less than 2 m deep. They have a silty and muddy sand substrate. Salinity and temperature data are summarized in Table 10.

Table 10 Environmental changes, Texas bays. The figures in parentheses give the size of the variation (data from Phleger and Lankford, 1957)

Date of observation	Upper lagoon		Lower lagoon	
	Temp. (°C)	Salinity (per mille)	Temp. °(C)	Salinity (per mille)
June, 1955	28.7–30.3 (1.6)	26.71–36.67 (9.96)	28.8–30.5 (1.7)	36.01–42.55 (6.54)
May, 1955	24.5–26.0 (1.5)	30.90–33.69 (2.79)	24.5–26.7 (3.2)	33.30–36.75 (3.45)
March, 1955	16.8–17.0	32.09–33.87 (incomplete data)	16.0–19.5 (3.5)	29.31–34.42 (5.11)
January, 1955	16.3–10.1 (2.8)	27.61–34.72 (7.11)	14.7–21.6 (6.9)	32.24–37.04 (2.80)
November, 1954	18.6–19.2 (0.6)	28.96–34.21 (5.25)	17.8–20.1 (2.3)	33.19–36.77 (3.58)
August, 1954	No data		No data	

During January, March and May, slightly hyposaline conditions prevail in the lower parts of the lagoons. In June, and probably throughout the summer, conditions are hypersaline. In the upper lagoon, slightly hyposaline conditions persist throughout the year.

Diversity is generally low, with α values from 1 to 4, but during January, March and May, 1955, some values rose to $\alpha = 8$ (Figure 26). On the triangular plots (Figure 27) there is a general dominance of Rotaliina with subsidiary Textulariina and Miliolina, but occasional samples plot close to the Miliolina corner. Standing crop is variable and reaches a maximum of 2608 per 10 cm², but most values are between 188 and 353 in the upper lagoon and 53 and 106 per 10 cm² in the lower lagoon.

The dominant species are *Ammonia beccarii* (Linné) (=*Streblus beccarii*

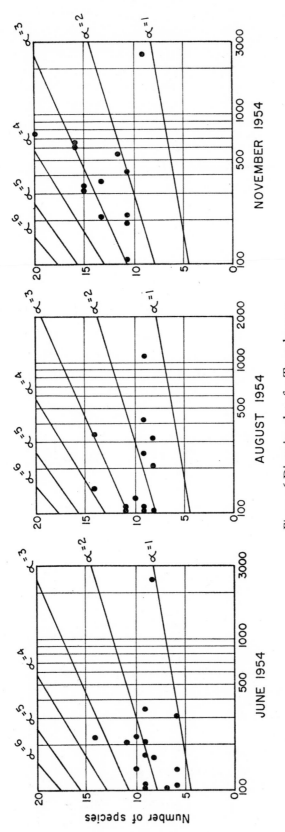

Figure 26 Diversity plots for Texas lagoons.

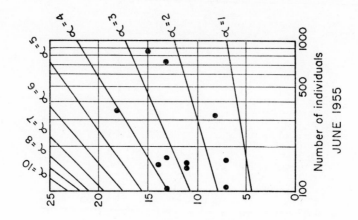

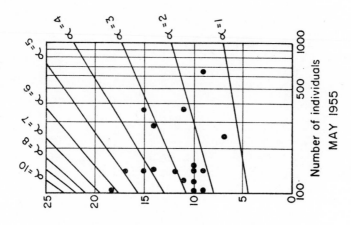

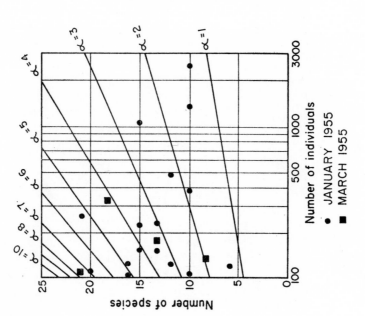

Figure 26—Continued.

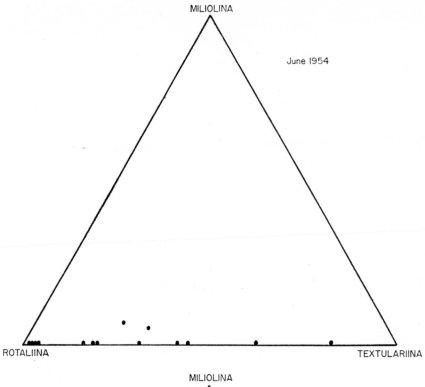

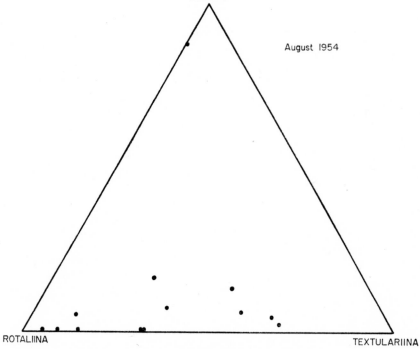

Figure 27 (Part 1) Triangular plots for Texas lagoons.

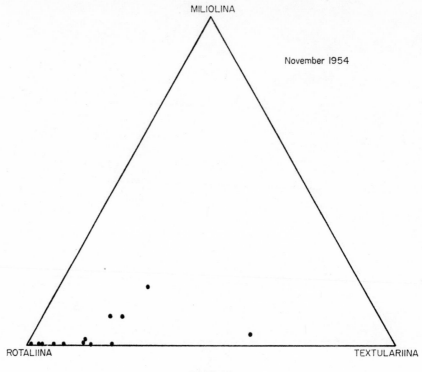

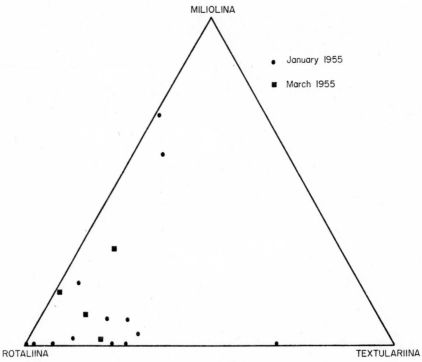

Figure 27 (Part 2).

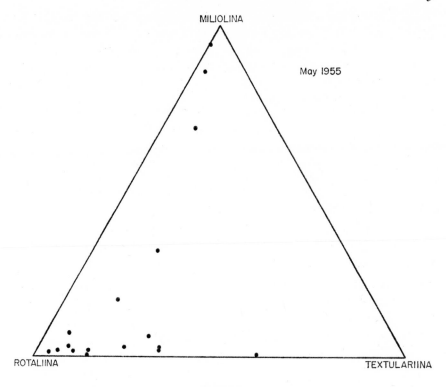

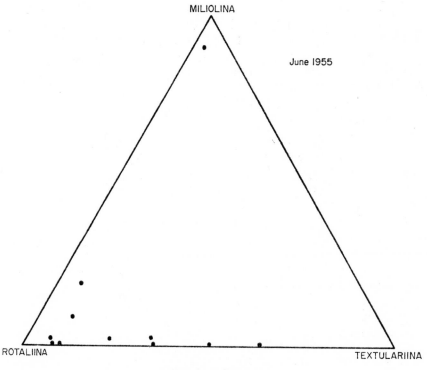

Figure 27 (Part 3).

A and B of Phleger and Lankford) in the lower lagoon and *Ammotium salsum* (Cushman and Brönnimann) (=*Ammobaculites salsus* of Phleger and Lankford) in the upper lagoon. Additional species are *Elphidium* spp., *Eponidella gardenislandensis* Akers (particularly in the upper lagoon) and, locally, *Quinqueloculina* spp.

Most of the fauna is comprised of typically hyposaline species. These have favourable conditions during January–May (hence the higher α values) but throughout the summer conditions are hypersaline and therefore adverse. The average standing crop is reduced and the α values are very low. By November the environment has returned to hyposaline conditions, but the fauna has not 'caught up' so that the α values and standing crops are still low. This seems to be an instance of a fauna not being in equilibrium with its environment (Murray, 1968b).

Hypersaline lagoons

As hypersaline water results from evaporation exceeding fresh-water runoff, hypersaline lagoons are a feature of subtropical and tropical areas. Two principal types occur:

  (a) those with carbonate sediment formed *in situ*, e.g. Trucial Coast of the Persian Gulf, Florida Keys, Florida Bay.
  (b) those with clastic sediments, Laguna Madre, Laguna Ojo de Liebre.

The Trucial Coast lagoon of Abu Dhabi is discussed in detail in Chapter 16 (special features of tropical carbonate sediments) so that only a brief summary is given here. The main environmental parameters are listed in Table 11. The α indices are low (see Figure 28). On the triangular plot

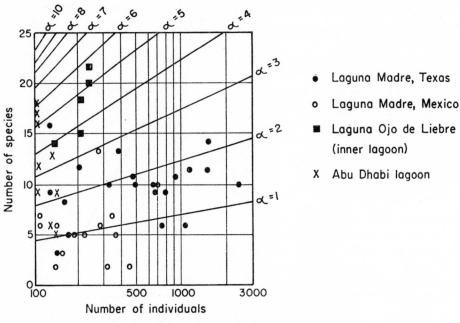

Figure 28 Diversity plot for hypersaline lagoons.

(Figure 29) the assemblages are dominated by Miliolina and to a lesser extent Rotaliina; the Textulariina are scarcely represented. The standing crop is low at 1–47 per 10 cm². During 1969 when the largest standing crop values were observed, the dominant species in the outer lagoon were *Peneroplis planatus* (Fichtel and Moll) associated with *Peneroplis pertusus* (Forskål), *Quinqueloculina* sp., *Triloculina* sp., *Miliolinella* sp., *Vertebralina striata*

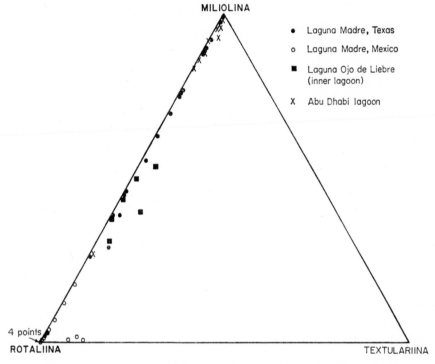

Figure 29 Triangular plots for hypersaline lagoons.

d'Orbigny, *Elphidium* spp. and *Ammonia beccarii* (Linné). The inner lagoon was dominated by *Peneroplis planatus* (Fichtel and Moll) and *Elphidium* cf. *E. advena* (Cushman). However, foraminiferids live only in association with sea-grass, seaweed or other algal growths. They are very rare on sediment surfaces in this environment (Murray, 1970b).

Brief notes on the assemblages of Coupon Bight, Florida Keys, have been given by Howard, Kissling and Lineback (1970). They record 55 species, of which 31 are Miliolacea. *Archaias* and *Quinqueloculina* are the dominant genera. The standing crop is 130–1670 per 10 cm². They distinguished six subenvironments, open bay, restricted bay, mangrove bay, bay mouth and tidal channel (all with a flora of *Thalassia* or algae) and the nearshore region. There are variations in the foraminiferal assemblages with *Archaias angulatus* (Fichtel and Moll) most abundant in the nearshore, restricted bay and tidal channel assemblages. Salinities range from 36.8–41.4 per mille and temperatures from 28.9 to 33.1 °C.

In Upper Florida Bay, Lynts (1962) recorded very low standing crop

Table 11 Summary of hypersaline lagoon data

	Laguna Madre, Texas, U.S.A.	Laguna Madre, Mexico	Laguna Ojo De Liebre, Baja California (Inner lagoon)	Abu Dhabi Lagoon, Trucial Coast, Persian Gulf	
Location					
Author	Phleger (1960c) Rusnak (1960)	Ayala-Castañares and Segura (1968)	Phleger and Ewing (1962)	Murray (1965a, 1970a, b)	
Sediment Type	Mixed sand, silt and clay away from barrier island	Silt, clay, evaporites	Fine sand	Carbonate, pellet, aragonite mud and shell debris	
Surface layer	Oxidized	—	Oxidized	Oxidized	
Disturbance	Water normally turbid due to wind induced waves	—	Small?	Periodically by storms / Normally by bioturbation	
				Outer lagoon	Inner lagoon
Salinity (per mille) Diurnal variation	—		—	—	—
Summer	Northern 40–80 / Southern 36–63	Up to 117	45–47	43–58	54–69
Autumn			44–46	—	—
Winter			42	43–52	50–72
Spring			—	43–50	48–54

			Average 1957 (1956)	*Outer lagoon*	*Inner lagoon*
Temperature (°C)					
Diurnal variation	—		—	—	—
Summer	—	Max. 40	26.8	32–36	34–37
Autumn	—	(brace for Summer–Spring)	20.0	—	—
Winter	—		19.5 (17.1)	23–25	18–24
Spring	—		—	23–26	22–24.8
Depth (m)	0–2.5 Average 1.3	0.7	0.5	0–3	0–1.5
Tides					
Number	None	None	Semidiurnal		Semidiurnal
Amplitude (m)			1–2.6		0.5–1
Foraminiferida Main species	Elphidium poeyanum; *E. galvestonensis; Quinqueloculina spp.; Triloculina oblonga; Triloculina obliquinoda; Ammonia beccarii	Ammonia beccarii; Locally Eponidella gardenislandensis; Palmerinella palmerae	Elphidium spp.; Quinqueloculina spp.; Q. laevigata; Q. limbata; Ammonia beccarii	Outer lagoon; Peneroplis planatus; P. pertusus; Quinqueloculina sp.; Triloculina sp.; Miliolinella sp.	Inner lagoon; Peneroplis planatus; Elphidium cf. E. advena
Fisher α indices	Northern 1–3.25; Southern 1–4.25	<1–3	4–6.25	2.5–6	1–2.5
Standing crop per 10 cm²	North 10–904, av. 200; South 0–2480, av. 800	0–668; Av. <100	7–257, av. 200–250	1–38	1–47

* Less common in southern part.

values: 1–52 per 10 cm². None of his samples can be used for α index or triangular plot. Previously Moore (1957) had also noted the sparse occurrence of living foraminiferids.

Laguna Madre, Texas, can be divided into northern and southern parts with salinities of 40–80 per mille and 36–63 per mille respectively (Table 11). It is an area of muddy and silty sands. Phleger (1960c) does not record the presence of any submarine vegetation, although Rusnak (1960) stated it to be common. The α values are generally from below 1 up to 3, with one value of 4.5 (Figure 28). On the triangular plot (Figure 29), all the points fall along the Miliolina–Rotaliina side. The standing crop values are high with an average of 200 per 10 cm² and a maximum of 2480 per 10 cm². The dominant species are listed in Table 11.

Laguna Madre, Mexico, is a very extreme environment with salinities up to 117 per mille and temperatures up to 40 °C. The average depth is 0.7 m, and during periods when the sand barrier island closes the channels to the sea, evaporation causes deposition of minerals such as halite and gypsum. The sediment is silt or clay away from the barrier island. No vegetation is recorded by Ayala–Castañares and Segura (1968). The standing crop values are generally low (only 13 out of 58 samples have greater than 100 per 10 cm²). Diversity is low, with α values from 1 to 3 (Figure 28). On the triangular plot (Figure 29) most of the samples lie close to the Rotaliina corner. The dominant species throughout is *Ammonia beccarii* (Linné) but locally *Eponidella gardenislandensis* Akers and *Palmerinella palmerae* Bermudez are common. The general rarity of Miliolacea is unusual for this type of lagoon.

On the Pacific coast of Baja California, the inner part of Laguna Ojo de Liebre is hypersaline (42–47 per mille, Phleger and Ewing, 1962) although temperatures are not particularly high. The α values are 4 to 6.5 (Figure 28), and on the triangular plot the points fall along the Miliolina–Rotaliina side. The main species are *Quinqueloculina* spp., *Elphidium* spp. and *Ammonia beccarii* (Linné).

The high standing-crop values (average 200–250 per 10 cm²) may be due to high primary production by phytoplankton. Values of carbon fixation given by Phleger and Ewing are 28–63 mg C/m³/day which compares with values of 1 for the Pacific 200 miles off Baja California, 2–6 for northern Gulf of Mexico and 21–45 for California Channel Islands.

Normal marine lagoons

Information on the known examples is listed in Table 12. Laguna Guerrero Negro and Laguna Ojo de Liebre have near normal salinities in their channels. The climate is warm temperate. The most abundant living foraminiferids are Miliolacea, particularly species of *Quinqueloculina*. The forms illustrated by Phleger and Ewing (1962, plate 1) are mainly smooth or striated. Additional common forms are species of *Elphidium* and *Glabratella*. The standing crop values are moderate.

The diversity indices (Figure 30) range from α = 4 to α = 11. On the triangular plot (Figure 31), the two lagoons occupy slightly different fields but both show dominance of Rotalina and/or Miliolina with subsidiary Textulariina (<15 per cent).

Table 12 Summary of normal marine lagoon data

Location	Laguna Guerrero Negro, Baja California	Laguna Ojo De Liebre, Baja California (Channel)	St Lucia
Author	Phleger and Ewing (1962) Phleger (1965b)	Phleger and Ewing (1962)	Schafer and Sen Gupta (1968)
Sediment			
Type	Fine sand	Fine sand	Muddy sands
Surface layer	Oxidized	Oxidized	Oxidized
Disturbance	Turbulent in channels	Intense turbulence in channels	
Salinity (per mille)			
Diurnal variation	Very small	—	—
Summer	35.3 at entrance	⎫	—
Autumn	—	⎪	—
Winter	34.7 at entrance	⎬ 34–35	—
Spring	—	⎭	34.0–35.5
Temperature (°C)		*Average 1957 (1956)*	
Diurnal variation	—	—	—
Summer	20.7–26.8	24.0	—
Autumn	18.7–20.0	18.7	—
Winter	14.8–19.5	18.1 (14.6)	—
Spring	—	—	26.0–27.7
Depth (m)	Channel 10 Inner lagoon 1–2	0–25	Av. 11
Tides			
Number	Semidiurnal	Semidiurnal	No data
Amplitude (m)	3 (springs) av. 13–16		
Foraminiferida			
Main species	Ammonia beccarii 'Bolivina' spp. Buliminella elegantissima Reophax nanus Rosalina columbiensis *Quinqueloculina laevigata *Q. limbata *Elphidium spp. *Glabratella spp.	Quinqueloculina spp. Q. laevigata Locally common 'Bolivina' spp. Elphidium spp. Rosalina columbiensis 'Rotalina' lomaensis versiformis	Ammonia beccarii
Fisher α indices	4–8	4–11	—
Standing crop per 10 cm²	60–778	105–1652, av. 100–200	1–150

* Main species.

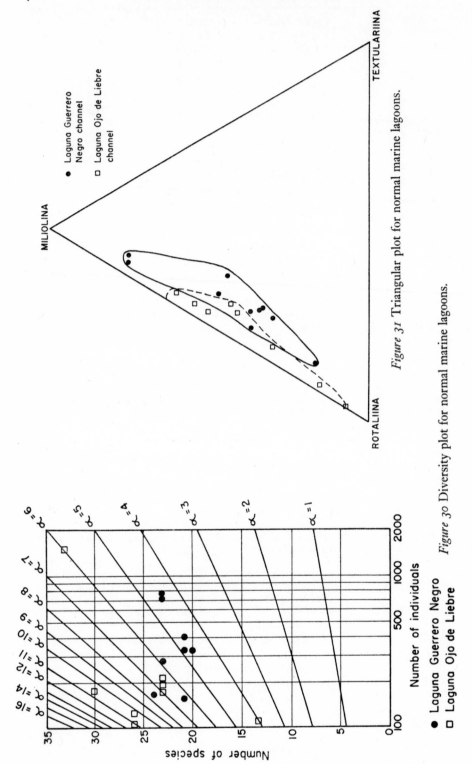

Figure 31 Triangular plot for normal marine lagoons.

Figure 30 Diversity plot for normal marine lagoons.

In a preliminary note on Port Castries Bay, St Lucia, Schafer and Sen Gupta (1968) reported the dominant superfamilies to be Rotaliacea, Buliminacea and Miliolacea. *Ammonia beccarii* (Linné) forms up to 38.4 per cent of the living assemblage (no other species were named although 46 were recorded). There are no data from which to plot diversity or the ratios of the three suborders.

Comparison of lagoons and estuaries

Estuaries and lagoons are all marginal marine environments where conditions are subject to big diurnal and seasonal changes. It is impossible for animals or plants to reach a state of equilibrium with such changing conditions. As a consequence, diversity is low and this is a characteristic feature of all lagoons (Figure 32). Most of the α values lie between 1 and 5 in

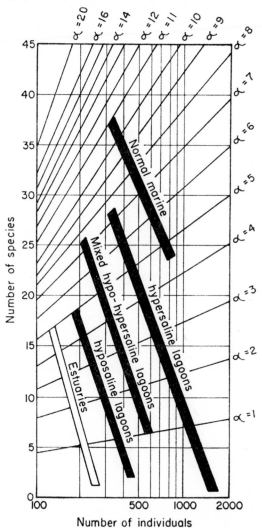

Figure 32 Summary of diversity data for lagoons and estuaries.

hyposaline and hypersaline environments. A few values of up to $\alpha = 7.5$ are encountered in hypersaline lagoons. Where normal marine conditions apply, the diversity range is higher and this reflects the more stable and favourable salinity conditions.

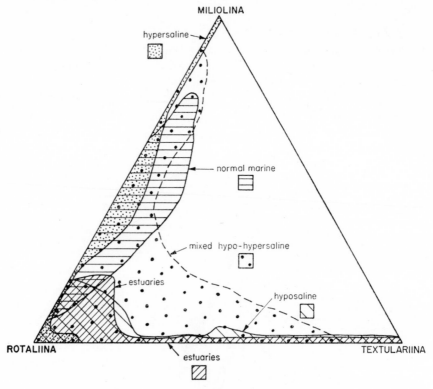

Figure 33 Summary triangular plot for lagoons and estuaries.

The summary triangular plot, Figure 33, shows different fields for the different environments. As in the α plot, estuaries and hyposaline lagoons cannot be differentiated and this is perfectly reasonable. However, hypersaline environments with their low abundance of Miliolina are clearly separable from hypersaline and normal marine lagoons. The latter two fields partially overlap. Normal marine lagoons are distinguished from hypersaline examples by having fewer Miliolina and more Textulariina. The Texas lagoons, showing an annual variation from hyposaline to hypersaline conditions, represent a compromise between these two fields.

Where Miliolina are present in hyposaline areas they are invariably close to the lagoon mouth and probably come under near-marine influences for a part of each day. Experiments show that *Quinqueloculina seminulum* (Linné) loses control of its pseudopodial activity when salinity is reduced to 30 per mille or less. This would prevent it from colonizing hyposaline areas and could account for the general absence of Miliolina from such environments (Murray, 1968b).

Consideration of the distribution of genera and species shows that *Ammotium*, *Miliammina fusca* (Brady) and *Protelphidium* are restricted to hyposaline environments (they occur in marshes as well as lagoons). Other common lagoon foraminiferids such as *Ammonia beccarii* (Linné) and *Elphidium* spp. occur in all types of lagoons (Figure 34). No species are confined to hypersaline lagoons; the same types are found also in normal marine examples.

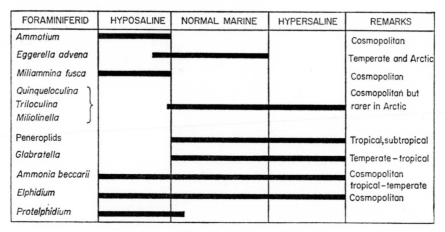

FORAMINIFERID	HYPOSALINE	NORMAL MARINE	HYPERSALINE	REMARKS
Ammotium				Cosmopolitan
Eggerella advena				Temperate and Arctic
Miliammina fusca				Cosmopolitan
Quinqueloculina ⎫ Triloculina ⎬ Miliolinella ⎭				Cosmopolitan but rarer in Arctic
Peneroplids				Tropical, subtropical
Glabratella				Temperate – tropical
Ammonia beccarii				Cosmopolitan tropical – temperate
Elphidium				Cosmopolitan
Protelphidium				

Figure 34 Distribution of lagoon genera.

Individual species show geographic restriction in their distribution. For hyposaline environments on the Atlantic seaboard of North America, a northern province extends northwards from Long Island Sound. It is characterized by cold-water species of *Elphidium* (particularly *E. clavatum*) and *Eggerella advena* (Cushman). Extending from Cape Cod southwards is a temperate province characterized by the presence of *Ammonia beccarii* (Linné) in addition to the cold-water forms. In the Gulf of Mexico, warm-water species of *Elphidium* occur with *Ammonia beccarii* (Linné) and *Ammotium*. *Miliammina fusca* (Brady) is cosmopolitan, but even it cannot withstand the severe conditions of the Hudson Bay estuaries.

Along the European seaboard *Ammonia beccarii* (Linné) extends as far as the entrance to the Baltic and it is absent from Bottsand Lagoon.

4. Deltas

'Modern deltas are the Recent plains which were formed at the margins of oceans, seas, bays and lakes by deposition of streamborne sediments at the mouths of streams.' (Rainwater, 1966, p. 3). Because of their great geological importance, many sedimentary studies of deltas have been made but the living foraminiferids of only two examples are known in detail.

MISSISSIPPI DELTA

This has been studied in greater detail than any other delta. Phleger (1955) and Lankford (1959) studied 266 samples from the eastern part. They recognized a series of intergrading environments ranging from the delta top to the open shelf (see Table 13). The interdistributary bay environment is shallow and hyposaline, and experiences a large annual temperature variation. The water is turbid due to suspended sediment. The fluvia-marine environment occupies the distributary channels of the river. During periods of heavy runoff, no sea water enters the channels. During periods of low runoff, there is a well developed salt wedge. The annual salinity variation is thus extreme. The Sound (i.e. lagoon) is shallow, hyposaline and less variable in temperature. The water is turbid. The deltaic marine environment includes the delta front platform and prodelta slope. Salinities are near normal for the open Gulf. The water is less turbid than that of the previous environments. Away from the immediate influence of the delta proper is the open shelf.

Phleger and Lankford placed the boundaries of these environments in slightly different positions. The author places the limit of the Sound along a line joining Breton Island and Baptiste Collette subdelta.

The standing crop of living foraminiferids is lowest in the Sound, fluvial marine and open shelf areas. High values prevail in the deltaic marine environment at the foot of the prodelta slope. Lankford found an average of 2500 per 10 cm^2 and a maximum of 8000. He noted that the foraminiferids are smaller than usual and that sedimentation is fast. He concluded that optimum conditions for growth and reproduction prevail so that maturity is reached quickly. Because the sediment accumulates so fast, the foraminiferids have to crawl up through the rain of detritus to keep near the sediment surface.

The diversity values show much the same range for all environments, $\alpha < 1$ to $\alpha = 5$. The only stable normal marine area is the open shelf (5 samples, Figure 35). On the triangular plot the majority of the samples from all the environments plot close to the Rotaliina corner. With the exception of five samples with 15–39 per cent Miliolina, all are in the field occupied by hyposaline lagoons. The exceptions include two sound, two open shelf and one deltaic marine samples (Figure 36).

Table 13 Data for the Mississippi Delta (based on Lankford, 1959 and Phleger, 1955)

	Interdistributary bay	Fluvial marine	Deltaic marine	Sound	Open shelf
Depth (m)	0.1-2	1.6-10	>10	1-10	0-120
Salinity (per mille)	1-10	1-32	34-36	18-36	<100
Temp. (°C)	0-38	8-29	19-30	15-30	17-31
Turbidity	Yes	Yes	Some	Yes	Small
Standing crop per 10 cm²	4-700	4-700	up to 8000	up to 500 generally <100	generally <100
α index	<1-3	1-3	1-5	1-5	2-4
Dominant species	Ammotium salsum Elphidium spp. Miliammina fusca Ammonia beccarii	Ammotium salsum Elphidium gunteri Palmerinella garden-islandensis	Nonionella opima Bolivina lowmani Buliminella cf. B. basendorfensis Epistominella vitrea	Ammotium salsum Elphidium gunteri Ammonia beccarii	No dominant sp.

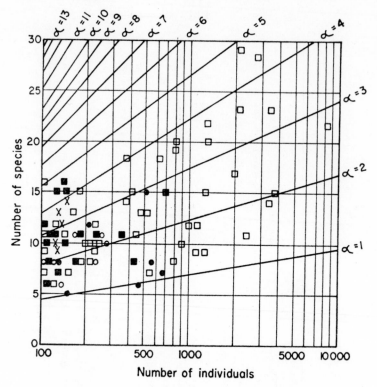

Figure 35 Diversity plot for the Mississippi Delta
(symbols as in Figure 36).

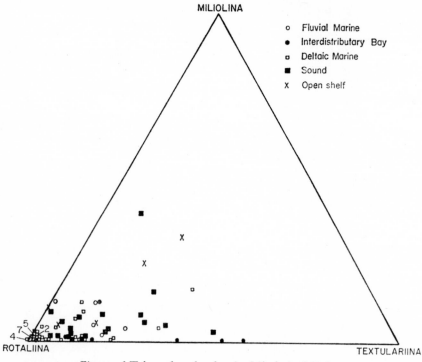

o   Fluvial Marine
•   Interdistributary Bay
◻   Deltaic Marine
■   Sound
x   Open shelf

Figure 36 Triangular plot for the Mississippi Delta.

The dominant species are listed in Table 13. The outstanding feature is the dominance of *Nonionella opima* Cushman in the deltaic marine environment.

EBRO DELTA, SPAIN

This delta is situated on the Mediterranean coast of Spain. It now receives only a small amount of runoff because most of the water is used for irrigation. Details of the environments are summarized in Table 14. The delta is fairly small. Sandy sediment forms the beaches and delta front platform (inshore), while muddy sediments occur on the prodelta slope (offshore).

Diversity is low in the lagoons where salinities sometimes are slightly hypersaline and temperatures are variable. In the deltaic marine and inshore delta flank areas, low diversities occur close to the shore and values increase with increasing depth. On the triangular plot, the inshore and offshore areas occupy separate fields (Figure 37). The inshore field is similar to that for

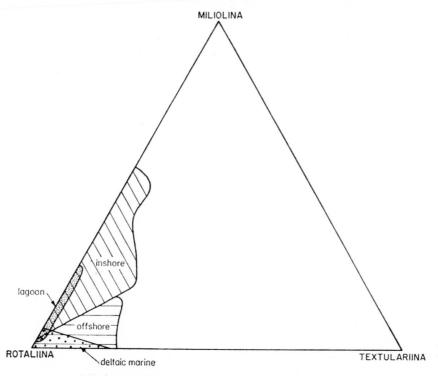

Figure 37 Triangular plot for the Ebro Delta, Spain.

normal marine lagoons. The offshore field is the same as that of shelf seas (Figure 102). Standing crop values are very low compared with those of the Mississippi Delta. This is probably related to the small input of nutrient-rich river water and hence the lower primary production.

The dominant species are listed in Table 14. In areas of muddy sediment, deltaic marine and offshore, *Nonionella opima* Cushman is the dominant species together with *Brizalina pseudopunctata* (Höglund) and *Bulimina aculeata* d'Orbigny. (Scrutton, *pers. comm.*)

Table 14 Data for the Ebro Delta (Scrutton pers. comm.)

	Lagoon	Delta marine	Delta flank	
			Inshore	Offshore
Depth (m)	0–10	0–>50	0–10	10–>50
Salinity (per mille)	>35 except near shore	37	37	37
Temperature (°C)	10–30	12–27	12–27	12–27
Turbidity	clear	turbid	clear	clear
Standing crop per 10 cm^2	3–140 average 45	16–234 average 30–90	10–100 average 30	27–326 average 100
α index	2–6	3–12	2–10	6–13
Dominant species	Ammonia beccarii Brizalina striatula B. pseudopunctata Hopkinsina pacifica atlantica Nonionella opima plus on weeds Quinqueloculina schlumbergeri	Shallow Ammonia beccarii Elphidium spp. Deeper Valvulineria complanata Nonionella opima Epistominella vitrea Bulimina aculeata Brizalina pseudopunctata	Eponides granulata Quinqueloculina schlumbergeri Asterigerinata sp. Nonion depressulus Ammonia beccarii	Nonionella opima Brizalina pseudopunctata Bulimina aculeata Valvulineria complanata

RHÔNE DELTA

Blanc-Vernet (1969) presented data on the occurrence of living and total assemblages along a profile of stations off the mouth of the Rhône. Figure 38 has been plotted from the living data. Station F_3 is off the river mouth

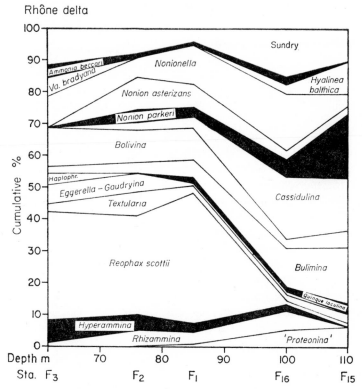

Figure 38 Cumulative histogram along a traverse from the nearshore (station F_3) to prodelta slope (station F_{15}) on the Rhône Delta (data from Blanc-Vernet, 1969).

and Station F_{15} is at the seaward end of the profile. A clear faunal break occurs between 85 and 100 m. At depths of less than 85 m the dominant species are:

Reophax scottii Chaster
Textularia spp.
Eggerella/Gaudryina
Bolivina sp.
Nonion asterizans (Fichtel and Moll)
Nonionella spp.

Textulariina form slightly more than 50 per cent and Rotaliina slightly less than 50 per cent of the living assemblage. The Miliolina are scarcely represented.

At depths greater than 85 m the Textulariina are much reduced (10–15 per cent) and the Rotaliina much increased. The dominant species are:

Bulimina spp.
Cassidulina spp.
Nonion parkeri le Calvez
Nonion spp.
Hyalinea balthica (Schröter)

The sediment is mud at all stations except F_{15} where the sands are believed to be remanié.

The occurrence of dead foraminiferids on the Rhône delta has been described by Kruit (1955).

SUMMARY AND CONCLUSIONS

The Mississippi delta has high runoff, and therefore nutrients are added to the sea and primary production is high (Phleger, 1964b). Fine-grained sediment is transported in large quantities so acretion is fast. Conditions are favourable for mud-loving forms like *Nonionella opima* so reproduction occurs regularly and standing crops are large. In the Ebro delta, runoff is low so primary production, standing crop and sedimentation are low.

No species are specific of deltas. *Nonionella opima* lives on muddy shelf sediments off the coast of California. Diversity and triangular plot data show a transition from shelf to hyposaline environments and this is really all a delta represents. The only unusual feature of deltas is their relatively fast rate of sediment build-up compared with that of shelf seas. Thus deltas do not form a distinct environment for foraminiferids but just an assemblage of environments (marsh, lagoon, inner and outer shelf) such as occur elsewhere.

5. Shelf Seas: Pacific Seaboard of North America

As this coastline is on the east side of the Pacific Ocean, the main direction of water movement is to the south. The Californian Current is a continuation of the Aleutian Current of the North Pacific. It is very wide and slow moving. To the south it joins the westward flowing North Equatorial Current. Sverdrup, Johnson and Fleming (1942, p. 724) state that in the spring and early summer, north-north-west winds off the Californian coast cause upwelling from March to July. Areas of upwelling are marked by lower temperatures and high nutrient concentrations, although the upwelled water is believed to arise from only about 200 m depth. During this time a countercurrent of equatorial water flows close to the coast at depths greater than 200 m.

At the end of the summer, upwelling stops and eddies develop which transport coastal water oceanwards and ocean water landwards. By the autumn a surface countercurrent, the Davidson Current, begins to flow to the north at least to latitude 48 °N. The countercurrent, below 200 m, continues during the winter months.

Because the Californian Current is a cold current surface, water temperatures are lowered. In August the temperature is around 15 °C as far south as the Mexican border. It rises to 20 °C at I. Cedros and 27 °C at the southern tip of Baja California (Figure 39). In February, the comparable temperatures are 12–13 °C, 18 °C and 21 °C (Figure 40). At 200 m the temperatures are 8–9 °C (Sverdrup, Johnson and Fleming, 1942). The surface salinities are close to 34 per mille.

The area off southern California is of great geological interest because of the narrow mainland shelf, the extensive deeper water shelves and the basins and troughs which cut them. All these features have been described in detail by Emery (1960).

Probably no part of the sea floor has been more intensively studied for its living foraminiferid assemblages than that off southern California. This, of course, reflects the interest shown by scientists from the Allan Hancock Foundation and Scripps Institute of Oceanography. The areas studied have been approximately delimited on Figure 41. For convenience they can be subdivided into the following categories:

(a) Continental shelf and slope adjacent to the mainland: Uchio (1960) off San Diego; Zalesny (1959) Santa Monica Bay; Bandy, Ingle and Resig (1964c) San Pedro Bay.

(b) Polluted areas of continental shelf: Bandy, Ingle and Resig (1965a) Santa Monica Bay; Bandy, Ingle and Resig (1964a) Palos Verdes;

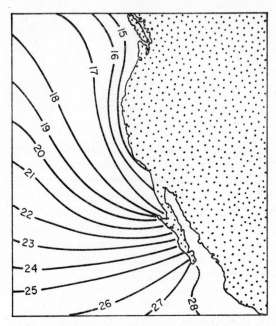

Figure 39 Surface isotherms for August (based on Sverdrup, Johnson and Fleming, 1942).

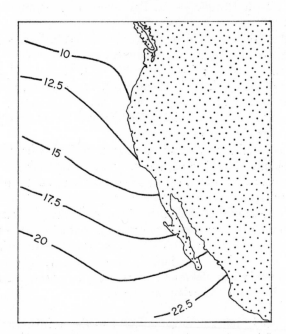

Figure 40 Surface isotherms for February (based on Sverdrup, Johnson and Fleming, 1942).

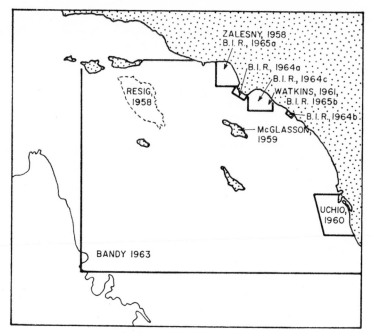

Figure 41 Map of the sea off southern California to
show the areas that have been studied. (B. I. R. =
Bandy, Ingle and Resig.)

Watkins (1961) and Bandy, Ingle and Resig (1965b) Orange County
sewer near Newport Beach; Bandy, Ingle and Resig (1964b) Laguna
Beach.

(c) Shelf around offshore island: McGlasson (1959) on Santa Catalina
   Island.

(d) Deep basins: Resig (1958), Bandy (1963).

SOUTHERN CALIFORNIA

Continental shelf and slope adjacent to the mainland

The most detailed study is that of Uchio (1960) on the San Diego area.
Here the mainland shelf is narrow, 6–15 km, with depths of up to 110 m
(Figure 42). Seaward of this the floor slopes down to the Loma Sea Valley
in the north, the Coronado Submarine Canyon in the centre and the San
Diego Trough in the south. The Coronado Bank separates the Loma Sea
Valley from the San Diego Trough north of Coronado Canyon. The complex pattern of sediment distribution has been described by Emery, Butcher,
Gould and Shepard (1952). Three broad groups have been recognized:
clastic, calcareous organic and mixed clastic-calcareous organic. In the
clastic group, silts are found in the deeper valley and trough areas, sandy silts
along the upper parts of the slopes, and fine to coarse sands on the mainland
shelf shallower than approximately 100 m. Calcareous organic sediments are
found on the Coronado Bank (foraminiferal sands) and around the Coronado

Islands (shell sand). Mixed sediments occur mainly off Point Loma. The
majority of samples considered by Uchio were from clastic sedimentary areas.

Hydrographic data quoted by Uchio were based on observations made from
1949–57. The salinity is close to 33 per mille at all depths throughout the
year. The oxygen minimum layer is about 600–700 m with values in the
range 0.25–0.48 cm³/l. Surface temperatures are 20 °C in August and

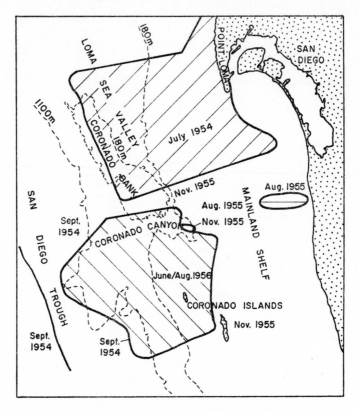

Figure 42 Map of the area studied by Uchio (1960)
off San Diego. The times and areas of sampling are
indicated.

September and 12 °C in January and February. A thermocline is present
throughout the year. The base is static at 75 m with temperatures of 9.3–
11.6 °C. The top varies seasonally from a winter maximum depth of 30 m
to nearly at the surface in August and the temperature from 13.1 °C to
19.3 °C.

Uchio (1960) collected samples on six different occasions during the
period 1954–56. The principal sampling times were July 1954 for the area
north of Coronado Canyon and June/August 1956 for areas to the south
(Figure 42).

The standing crop values are very high. In Figure 43 the data are generalized into three categories: populations of less than 500 per 10 cm², high
values of 500–1000 per 10 cm², and very high values in excess of 1000 per

10 cm². North of Coronado Canyon there is no clear correlation between standing crop and depth. Values greater than 1000 per 10 cm² are irregularly distributed and generally comprise single samples. In Coronado Canyon the very high values at the canyon head are clearly shown. The greater part of the Canyon has values of 500–1000 but in the lower part where the valley

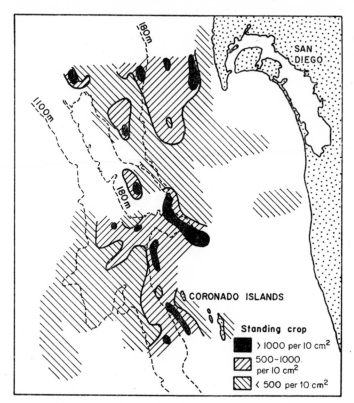

Figure 43 Generalized distribution of standing crop values (individuals per 10 cm²; based on data in Uchio, 1960).

opens on to the alluvial fan, values drop to less than 500 per 10 cm². On the shelf and slope south of the canyon values greater than 1000 per 10 cm² are discontinuously distributed at depths of approximately 100 m. On either side there is a zone of 500–1000 and then in shallower (<75 m) and deeper waters values drop to less than 500 per 10 cm². Comparatively low values of 100–250 per 10 cm² are typical of the San Diego Trough at depths of up to 1200 m.

Uchio compared standing crop size with sediment type. The coarse sands have low values regardless of depth (approximate average of 146–149 per 10 cm²). Fine sand averages 578 per 10 cm² but highest values are attained in the depth range 75–830 m. Silts give similar values with an average of 568 per 10 cm². Again highest values are attained in the depth range 37–550 m. Clayey silts average 536 per 10 cm², but there is no clear depth zone of higher abundance although below 1000 m in the San Diego

Trough values are constantly low. Low temperatures and low oxygen may be contributory factors here. Foraminiferal sands and silts have an average value of 629 per 10 cm².

In summary, there is clearly a correlation between low standing crops and coarse sediment, but fine sands, silts, clayey silts and foraminiferal sands all have similar average values and it therefore seems unlikely that sediment alone is a major factor in controlling standing crop throughout most of the area. Further, there is little evidence of big seasonal differences related to the time of collection of the samples.

In passing from shallow to deeper water, Uchio noted that the number of species generally increases. To the north of the canyon at depths greater than 180 m, the number of species decreases again, while to the south there is a second maximum in the depth zone 730–1100 m. He further commented that 'there is no positive correlation of areal distribution of the number of species with that of the size of the living population, the total population, or the L/T ratio'.

The Fisher α indices have been plotted both as graphs (Figures 44–47) and as a map (Figure 48). They have been summarized in Table 15. The

Table 15 Summary of Fisher α diversity indices in the depth zones proposed by Uchio (1960)

Depth (m)	July 1954	Aug. 1955 Aug. 1956	Sept. 1954 Nov. 1955 June 1956	Overall	General range
		Range of α values			
0–24	<2	2–4		<2–4	2–4
24–82	3.5–12	8.5–14		3.5–14	5–12
82–180	7.5–16	6–14.5		6–16	7.5–16
180–460	4.5–14.5	4.5–11.5	6–16	4.5–16	6–16
460–640		11.5–20	4–12.5	4–20	11–15
640–825		10–13.5	10.5	10–13.5	10–14
825–1190	10.5	9.5–19	8.5–19	8.5–19	9–19

area to the north of Coronado Canyon shows a progression from low α values in the nearshore zone through moderate α values (5–10) to high α values (10–15) away from the shore. However, in the very north there are both high and moderate values in the Loma Sea Valley. Coronado Canyon is an area of moderate α values (5–10) in the landward part and high α values (10–15) in the lower part. South of the canyon there is a progression from low to high values with increasing depth. On the slope and alluvial fan there are scattered low and high values but these do not seriously alter the regional picture. In the San Diego Trough some α values exceed 15.

In Figures 44–46 the values have been subdivided according to the time of sampling and the depth. It is difficult to know whether the differences observed are related primarily to true differences between the areas sampled or to the different times of sampling. On the whole it seems most likely that

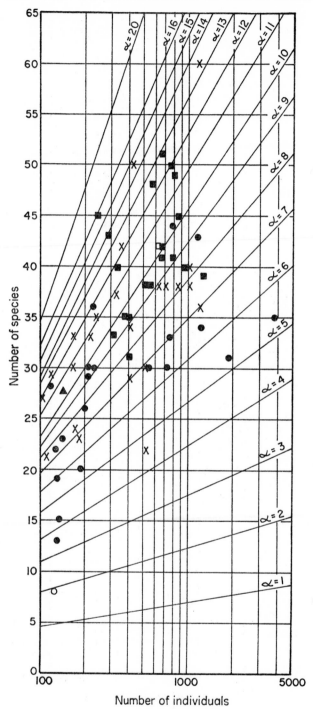

Figure 44 Fisher α diversity indices for the area north
of Coronado Canyon, sampled July 1954. o = 0–24 m,
● = 24–82 m, ■ = 82–180 m, × = 180–460 m,
+ = 460–640 m, □ = 640–825 m, ▲ = 825–1190 m.

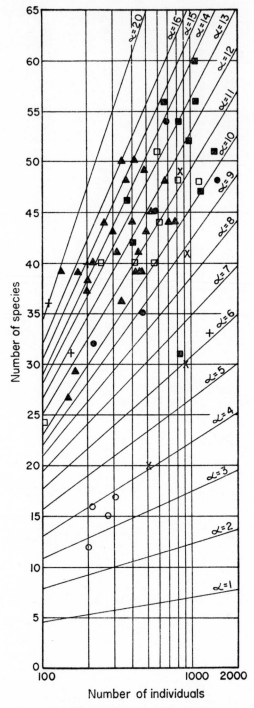

Figure 45 Fisher α diversity indices for the area of Coronado Canyon and to the south, sampled August 1955 and August 1956. Symbols as on Figure 44.

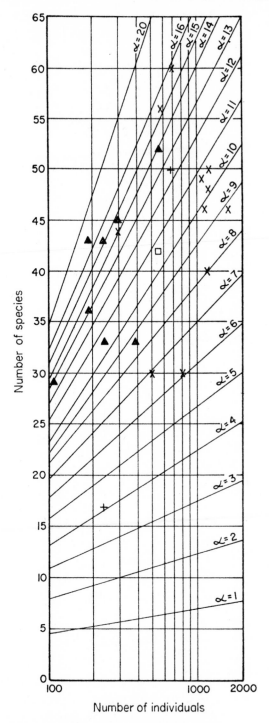

Figure 46 Fisher α diversity indices for various samples collected during September 1954, November 1955 and June 1956. Symbols as on Figure 44.

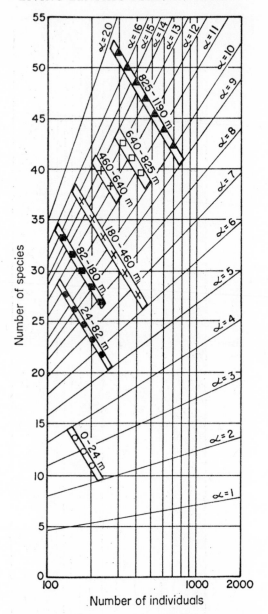

Figure 47 Summary of the general range of the Fisher α indices for the depth zones proposed by Uchio. Symbols as on Figure 44.

the former is true, as the area north of the canyon seems to be different in many respects from that to the south. The summary table shows that except for the low α values of the 0–24 m zone, all other depth zones proposed by Uchio are not clearly defined by overall differences of diversity. However, if the general range of α values is taken (i.e. excluding the few very high and very low values from each depth zone) a more definite picture emerges

(Figure 47). From this it can be stated that α values less than 4 indicate water of 0.24 m depth, α values greater than 4 indicate depths greater than 24 m, α values greater than 12 indicate a depth of more than 82 m and values greater than 16 indicate depths of more than 825 m.

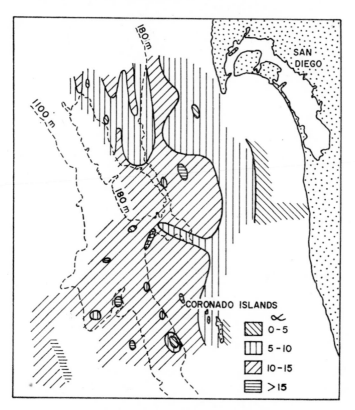

Figure 48 Map of Fisher α indices.

A triangular plot of all samples of more than 100 individuals (Figure 49) shows two fields. The great majority of samples, from all seasons and all depths, plot in a restricted field with 10–65 per cent Textulariina, 35–90 per cent Rotaliina and less than 3 per cent Miliolina. The second field generally has 3–11 per cent Miliolina, 3–39 per cent Textulariina and 58–95 per cent Rotaliina. All these latter stations are from depths of less than 55 m and most are from a belt of mixed clastic and calcareous organic sediments extending south from Point Loma. One unusual sample collected in August 1955 has 55 per cent Miliolina, 2 per cent Textulariina and 43 per cent Rotaliina. This is from an area of gravel and boulders off the Tia Juana River. It is probable that algae grow here and have created the environmental conditions necessary for such a high Miliolina concentration.

Uchio recognized a series of depth zones as follows: 0–24 m (0–13 fms, 24–82 m (13–45 fms), 82–180 m (45–100 fms), 180–460 m (100–250 fms), 460–640 m (250–350 fms), 640–825 m (350–450 fms), 825–1190 m (490–650 fms). The following tables list the species which consistently form

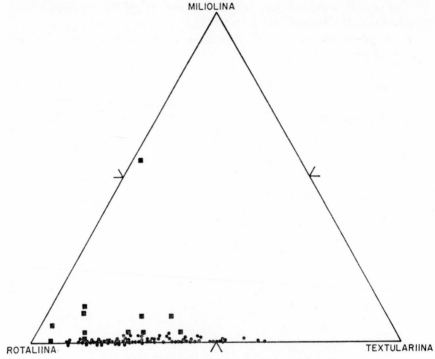

Figure 49 Triangular plot of samples from the San
Diego area. *Key:* ● elastic sediment, ■ mixed
elastic and calcereous sediment shallower than
55 m.

more than 5 per cent of the living assemblage in each depth zone (*indicates
that the species occurs over a great depth range; d = dominant species;
data from Uchio, 1960, text figures 4–9):

0–24 m
  On the fine sediment
    Elphidium spp.
    Buliminella elegantissima (d'Orbigny)
    Miliolids
    Nonionella basispinata (Cushman and Moyer)

  On coarse sand
    Rosalina spp.
    Neoconorbina terquemi (Rzehak)
    **Cibicides fletcheri* Galloway and Wissler
    Miliolids

The base of this zone is thought to coincide with the base of the turbulent
zone.

24–82 m
    Elphidium spp.
    Bolivina acutula Bandy

Cassidulina tortuosa Cushman and Hughes
Buliminella elegantissima (d'Orbigny)
Cibicides fletcheri Galloway and Wissler
Buccella angulata Uchio
Hanzawaia nitidula (Bandy)
Trifarina angulosa (Williamson) (=*Angulogerina angulosa* of Uchio)
d**Nonionella basispinata* (Cushman and Moyer)
d**Alveolophragmium columbiense* (Cushman)
**Fursenkoina sandiegoensis* (Uchio) (=*Virgulina sandiegoensis* of Uchio)
**Reophax scorpiurus* Montfort
**Cassidulina subglobosa* (Brady)
d*Reophax gracilis (Kiaer)
**Cassidulina depressa* Asano and Nakamura
**Trochammina kellettae* Thalmann
d**Nonionella stella* Cushman and Moyer
**Bolivina pacifica* Cushman and McCulloch
**Globobulimina pacifica* Cushman
*Goësella flintii Cushman

The base of this zone corresponds with the base of the seasonal thermocline. The species of this depth zone are fairly eurythermal.

82–180 m

Buccella angulata Uchio
Bulimina denudata Cushman and Parker
Cibicides mckannai Galloway and Wissler
Trifarina angulosa (Williamson) (=*Angulogerina angulosa* of Uchio)
d*Bolivina acuminata* Natland
**Alveolophragmium columbiense* (Cushman)
**Reophax scorpiurus* Montfort
**Cassidulina subglobosa* Brady
d*Reophax gracilis (Kiaer)
**Cassidulina depressa* Asano and Nakamura
**Arenoparrella oceanica* Uchio
**Alliatina primativa* (Cushman and McCulloch)
**Trochammina kellettae* Thalmann
d**Nonionella stella* Cushman and Moyer
d**Bolivina pacifica* Cushman and McCulloch
**Globobulimina pacifica* Cushman
d**Chilostomella ovoidea* Reuss
**Fursenkoina apertura* (Uchio) (=*Virgulina apertura* of Uchio)

The lower boundary of this depth zone may represent the boundary between the southward flowing Californian current and a deeper northward flowing current. Uchio points out that species found shallower than 180 m also occur off British Columbia, Canada. Species from deeper than 180 m are known also from deep water off South America.

180–460 m

Bulimina denudata Cushman and Parker
Gaudryina arenaria Galloway and Wissler
Fursenkoina delicatula (Uchio) (=*Virgulina delicatula* of Uchio)
*Cassidulina subglobosa Brady
d*Reophax gracilis (Kiaer)
**Arenoparrella oceanica* Uchio
**Textularia sandiegoensis* Uchio
**Trochammina pacifica* Cushman
**Alliatina primitiva* (Cushman and McCulloch)
**Trochammina kellettae* Thalmann
**Nonionella stella* Cushman and Moyer
d**Bolivina pacifica* Cushman and McCulloch
**Globobulimina pacifica* Cushman
d*Goësella flintii Cushman
**Fursenkoina apertura* (Uchio) (=*Virgulina apertura* of Uchio)
**Reophax micaceous* Earland
**Fursenkoina seminuda* (Natland) (=*Virgulina seminuda* of Uchio)
**Bolivina subargentea* Uchio

The lower boundary is possibly equivalent to the base of the permanent thermocline.

460–640 m

Trochammina chitinosa Uchio
**Textularia sandiegoensis* Uchio
**Trochammina pacifica* Cushman
**Alliatina primitiva* (Cushman and McCulloch)
**Trochammina kellettae* Thalmann
d**Nonionella stella* Cushman and Moyer
d**Bolivina pacifica* Cushman and McCulloch
**Globobulimina pacifica* Cushman
*Goësella flintii Cushman
**Chilostomella ovoidea* Reuss
**Fursenkoina apertura* (Uchio) (=*Virgulina apertura* of Uchio)
**Eponides leviculus* (Resig)
*Bolivina spissa Cushman
**Fursenkoina seminuda* (Natland) (=*Virgulina seminuda* of Uchio)
d**Bolivina subargentea* Uchio
**Cassidulina delicata* Cushman
**Loxostomum pseudobeyrichi* (Cushman)

The lower boundary of this depth zone is roughly equivalent to the oxygen minimum layer.

640–825 m

Epistominella smithi (R. E. and K. Stewart)
d**Bolivina pacifica* Cushman and McCulloch

Globobulimina pacifica Cushman
Chilostomella ovoidea Reuss
d**Fursenkoina apertura* (Uchio) (=*Virgulina apertura* of Uchio)
Eponides leviculus (Resig)
d*Bolivina spissa Cushman
Fursenkoina seminuda (Natland) (=*Virgulina seminuda* of Uchio)
Cassidulina delicata Cushman
Loxostomum pseudobeyrichi (Cushman)

825–1190 m

Valvulineria glabra* Cushman
Trochammina globigeriniformis* Parker and Jones
Uvigerina auberiana* d'Orbigny
Gyroidina io* Resig
Cibicides spiralis* Natland
Cassidulina subcarinata Uchio
Trochammina kellettae Thalmann
Nonionella stella Cushman and Moyer
d**Bolivina pacifica* Cushman and McCulloch
Globobulimina pacifica Cushman
Chilostomella ovoidea Reuss
d**Fursenkoina apertura* (Uchio) (=*Virgulina apertura* of Uchio)
Eponides leviculus (Resig)
Glomospira gordialis (Jones and Parker)
d*Bolivina spissa Cushman
Fursenkoina seminuda (Natland) (=*Virgulina seminuda* of Uchio)
Cibicides phlegeri Uchio
Valvulineria araucana (d'Orbigny)
Cassidulina delicata Cushman
Fursenkoina complanata (Egger) (=*Virgulina complanata* of Uchio)

A general discussion on the validity of depth zones is given on pages 168–72. To illustrate some of the faunal changes encountered in passing from the land to the deeper offshore areas, cumulative histograms have been prepared. The traverse in Figure 50 crosses the northern sampling area from the mainland shelf on the right to the Loma Sea Valley on the left. The depth boundaries recognized by Uchio at 82 and 180 m are indicated by arrows. On this traverse the most conspicuous break in the species distribution is at station 53: several deeper-water species die out and shallow ones start. This is emphasized by the drop in the similarity index from 60–88 per cent on the mainland shelf to 38 per cent between stations 53 and 54. In the Loma Sea Valley similarity is moderate between stations 54 and 59, but beyond this the low values indicate a further faunal change.

The same breaks can be seen on Figure 51. All similarity values of less than 50 per cent between adjacent samples have been delimited. In the area north of Coronado Canyon the variation in depth between adjacent samples is small. As a consequence, along each traverse there are several samples in each depth zone. These samples show similarity greater than 50 per cent and commonly 60–70 per cent. At the faunal breaks similarity between samples on either side of the break drops to less than 50 per cent and is commonly 30–40 per cent.

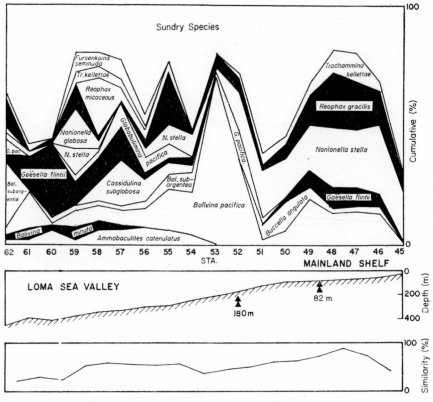

Figure 50 Cumulative histogram along a traverse from Loma Sea Valley to the mainland shelf (based on data from Uchio, 1960).

In the Coronado Canyon and to the south, the steep slopes cause big depth differences between adjacent samples. As a consequence each sample is in a different depth zone so that there are many faunal breaks and the pattern is more confused.

Santa Monica Bay

Zalesny (1959) studied the mainland shelf, slope and Redondo Canyon in Santa Monica Bay. Unfortunately much of his paper concerns total assemblages, and the only area for which living assemblage data are available is Redondo Canyon. Here depths vary from 66 m (36 fms) to 700 m (360 fms). Temperature varies from 10–11 °C in the shallow water to 5 °C at depth. Salinity is generally close to 34 per mille. Sediments throughout the canyon are silts.

A cumulative histogram of the traverse down the Redondo Canyon (Figure 52) shows marked faunal changes. At depths of 66–350 m, the domi-

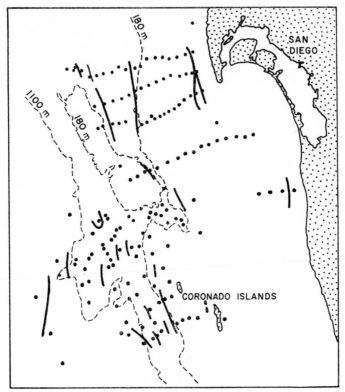

Figure 51 Map of faunal breaks revealed by similarity indices of less than 50 per cent.

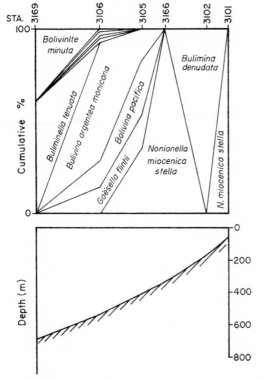

Figure 52 Cumulative histogram along a traverse down the Redondo Canyon (redrawn from Zalesny, 1959).

nant species are *Nonionella miocenica stella* Cushman and Moyer and *Bulimina denudata* Cushman and Parker. From 350 to 550 m *Nonionella miocenica stella* decreases in abundance and both species are replaced by *Goësella flintii* Cushman, *Bolivina pacifica* Cushman and McCulloch and *Bolivina argentea monicana* Zalesny. Deeper than 550 m the dominant species are *Buliminella tenuata* Cushman and *Bolivinita minuta* (Natland). Zalesny notes that for the most species the depth range of dead individuals is much greater than that of the living.

San Pedro Bay

The area sampled by Bandy, Ingle and Resig (1964c) varies in depth from 0 to 150 m. Only a thin veneer of sediment covers most of the shelf. Silty sand, sandy silt and sand are the common sediment types.

The number of living foraminiferids per gramme of sediment is 1–10. This figure cannot easily be related to standing crop per 10 cm². No tables of

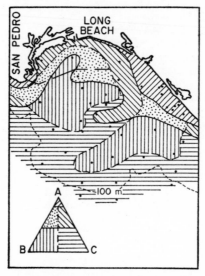

Figure 53 Distribution of the three faunal groups in San Pedro Bay. A = *Buliminella elegantissima* (d'Orbigny) group, B = *Trochammina pacifica* Cushman group, C = *Bulimina marginata denudata* Cushman and Parker group (redrawn from Bandy, Ingle and Resig, 1964c).

data were provided by the authors so it is not easy to re-interpret their results. Three bathymetric species groups were selected on previous knowledge:

1. *Buliminella elegantissima* group   0–150 m
   Buccella frigida (Cushman)
   Buliminella elegantissima (d'Orbigny)
   Fissurina lucida (Williamson)

2. *Bulimina marginata denudata* group   20–120 m
    Bulimina affinis d'Orbigny
    Bulimina marginata denudata Cushman and Parker
    Nonionella scapha basispinata (Cushman and Moyer)
    Nonionella miocenica stella Cushman and Moyer
3. *Trochammina pacifica* group   20–70 m
    Eggerella advena (Cushman)
    Gandryina arenaria Galloway and Wissler
    Goësella flintii Cushman
    Haplophragmoides advenum (Cushman)
    Reophax communis Lacroix
    Reophax scorpiurus Montfort
    Trochammina pacifica Cushman

The *Buliminella elegantissima* group is dominant in the nearshore areas, *Trochammina pacifica* group in the central shelf and the *Bulimina marginata denudata* group in the outer shelf and upper bathyal areas (Figure 53). The authors concluded that there is no correlation between the organic output of the sediment and the number of living foraminiferids.

The same basic faunal groups have been recognized on adjacent areas of the mainland shelf by Bandy, Ingle and Resig (1964a, b, 1965a, b). At depths shallower than 10 m and extending into the intertidal zone, an assemblage comprising *Rotorbinella lomaensis* (Bandy), *Cibicides fletcheri* Galloway and Wissler and miliolids is normally present.

Shelf around an offshore island

The shelf around Santa Catalina Island was studied by McGlasson (1959). The edge of the shelf is at a depth of approximately 85 m. To the north-east

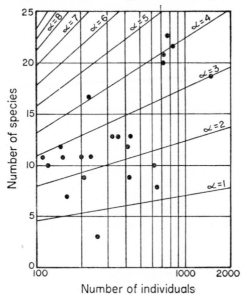

Figure 54 Fisher α diversity indices for the shelf around Santa Catalina Island.

and south-west are deep basins, the San Pedro and Catalina Basins respectively. Salinity is close to 34 per mille. Temperatures vary from 20 °C at the surface to a minimum of 8 °C at 200 m. The annual variation is 7 °C at the surface and 2 °C or less at depths greater than 50 m. There is considerable variation of sediment type, from gravel through silts to calcareous sand.

McGlasson's samples were of unknown area so no standing crop information is available. The diversity indices range from values below 1 up to 4.5 for those samples with more than 100 living individuals (Figure 54). On a triangular plot all the samples show a dominance of Rotaliina (Figure 55).

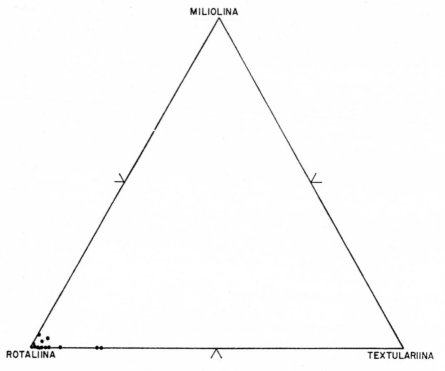

Figure 55 Triangular plot of shelf samples, Santa Catalina Island.

A considerable difference between the living and dead representatives in the depth zones was noted by McGlasson. The living forms are arranged in order of decreasing abundance:

0–37 m depth, 11.1–20.6 °C temperature
   Elphidium rugulosum Cushman and Wickenden
   Nonionella basispinata (Cushman and Moyer)
   Textularia articulata d'Orbigny
37–65 m depth, 9.6–14.2 °C temperature
   Bolivina quadrata Cushman and McCulloch
   Nonionella basispinata (Cushman and Moyer)
   Bolivina compacta Sidebottom

65–180 m + depth, 8.0–11.1 °C temperature
Bolivina pacifica Cushman and McCulloch
Cancris sagra d'Orbigny

In addition to these species *Bolivina acuminata* Natland is generally fairly common throughout the three depth zones.

Polluted areas of continental shelf

Watkins (1961) carried out the first study of the effects of sewer pollution on benthic foraminiferids. He sampled the area around the Orange County sewer between Huntingdon and Newport Beaches. Although he stained his material with rose Bengal, most of his discussion related to total assemblages. *Eggerella advena* (Cushman) normally lives in the depth zone 30–80 m, but around the outfall this species is abundant in much shallower water. The same is true of *Alveolophragmium columbiensis* (Cushman) and *Trochammina pacifica* Cushman. Size measurements of *Buliminella elegantissima* (d'Orbigny) and *Trochammina pacifica* Cushman show no effects introduced by the sewerage. However, *Elphidium spinatum* Cushman and Valentine and *Nonionella basispinata* (Cushman and Moyer) both develop abnormalities in the area of discharge.

Further studies by Bandy, Ingle and Resig (1965b) have shown that the number of living individuals at the point of discharge is low, but there is a surrounding aureole of high abundance (ten individuals per gram of sediment). These authors dispute Watkin's assertion that agglutinated species are more abundant near the outfall. Instead they find that hyaline foraminiferids are more than eight times abundant in the aureole than are miliolids or textularians. They suggest that the high abundance of dead agglutinated individuals may be due to postmortem solution of calcareous tests.

Buliminella elegantissima (d.'Orbigny) is the dominant species in the discharge aureole (>50 per cent of the living assemblage). *Trochammina pacifica* Cushman and *Eggerella advena* (Cushman) together comprise about 10 per cent of the living assemblage. Bandy *et al.* suggest that their main affinity is with silty sand rather than with sewage.

In Santa Monica Bay, the Hyperion sewage outfall of Los Angeles is one of the largest in southern California. The discharge is 1×10^9 litres/day. At present there are three discharge points.

Zalesny (1959) noted that *Trochammina pacifica* Cushman is most abundant around the Hyperion outfall. Bandy, Ingle and Resig (1965a) investigated the effects of the effluent on the foraminiferid distributions. The number of living specimens per gram of sediment is twice as great as the mainland shelf in the area of sewage outfall. The highest values are found not at the outfall proper, but downcurrent. Here the nutrients are used by the phytoplankton, and these in turn act as food for organisms such as foraminiferids.

Agglutinated species are 5–20 times more abundant than normal in the region 2–4 km downstream from the outfall discharge. *Eggerella advena* (Cushman) is 5–300 times as abundant; *Trochammina pacifica* Cushman is

2–1600 times as abundant and makes up 10 per cent of the living assemblage on sandy and silty sediments in the near-outfall area. Among the hyaline species, *Buliminella elegantissima* (d'Orbigny) is 5–500 times as abundant and makes up 30–50 per cent of the live assemblage. *Bulimina marginata denudata* Cushman and Parker is 10–900 times as abundant on silty sands and sandy silts, with nitrogen values of 0.3–0.5 per cent, and it makes up 50–100 per cent of the living assemblage.

Species which are adversely affected by pollution include *Nonionella miocenica stella* Cushman and Moyer and *Nonionella scapha basispinata* (Cushman and Moyer).

Around the Laguna Beach outfall, Bandy, Ingle and Resig (1964b) found living foraminiferids to be twice as abundant as elsewhere on the shelf. Hyaline foraminiferids are eight times as abundant as all other groups on the shelf. *Buliminella elegantissima* (d'Orbigny) is the dominant living species near the outfall, and is abundant elsewhere on the shelf too. In the immediate vicinity of the outfall, *Bolivina vaughani* Natland is one of the few living species. *Nonionella scapha basispinata* (Cushman and Moyer) is adversely affected by the sewage. Other species occurring in reduced abundance within 500 m of the discharge point include *Buccella frigida* (Cushman), *Cibicides fletcheri* Galloway and Wissler, *Discorbis columbiensis* Cushman, *Elphidium poeyanum tumidum* Natland, *Quinqueloculina* spp., and *Rotorbinella lomaesis* (Bandy).

Los Angeles County outfall opens on to the shelf off the Palos Verdes Hills. Bandy, Ingle and Resig (1964a) found a clear correlation between the occurrence of living foraminiferids and the primary sewage field. Along an elongate zone landwards of the outfall, living individuals are absent; this is the zone of suspended sewage. Away from this there is an increase in abundance and in the number of species. As in other polluted areas, agglutinated specimens are dominant in the dead assemblage, although in the living assemblage the Rotaliina are up to eight times more abundant. In particular the *Buliminella elegantissima* (d'Orbigny) group is dominant on the inner shelf and under the inner half of the sewage field. This includes *Buliminella elegantissima* (d'Orbigny), *Bolivina vaughani* Natland and *Buccella frigida* (Cushman). A second group comprising *Bulimina marginata denudata* Cushman and Parker, *Nonionella miocenica stella* Cushman and Moyer and *Nonionella scapha basispinata* (Cushman and Moyer) is abundant on the central and outer shelf, but under the outer part of the sewage field only the first species is dominant. The two *Nonionella* species are adversely affected by the sewage.

The sewer outfalls already described vary considerably in the size of their discharge. Bandy, Ingle and Resig (1964a, b, 1965a, b) quote the following rates:

Hyperion outfall	1×10^9 litres/day
Los Angeles County outfall	1×10^9 litres/day
Orange County outfall	284×10^6 litres/day
Laguna Beach outfall	13×10^6 litres/day

Thus the Hyperion and Los Angeles County outfalls are the same, and each is five times that of the Orange County outfall and 275 times the Laguna Beach outfall. It is therefore interesting to observe the same effects of pollution on living foraminiferids at each sewer.

Beneath the suspended sewage field there are few or no living foraminiferids, but away from this, either as an aureole or as a downcurrent field, abundance of living individuals increases to twice or more than that of the adjacent unpolluted shelf. The abundance of individual species may increase by as much as 500 times. All the sewers discharge at a depth of less than 100 m. On the unpolluted shelf, three groups of species characterize this depth zone:

1. *Buliminella elegantissima* group   0–150 m
    Bolivina vaughani Natland
    Buccella frigida (Cushman)
    Buliminella elegantissima (d'Orbigny)

2. *Bulimina marginata denudata* group   20–120 m
    Bulimina marginata denudata Cushman and Parker
    Nonionella miocenica stella Cushman and Moyer
    Nonionella scapha basispinata (Cushman and Moyer)

3. *Trochammina pacifica* group   20–70 m
    Eggerella advena (Cushman)
    Gaudryina arenaria Galloway and Wissler
    Goësella flintii Cushman
    Haplophragmoides advenum Cushman
    Reophax communis Lacroix
    Reophax scorpiurus Montfort
    Trochammina pacifica Cushman

In the sewage aureole each group shows an increase in abundance although not all the individual species are equally favoured. In particular *Buliminella elegantissima* (d'Orbigny) and *Bulimina marginata denudata* Cushman and Parker show a dramatic increase. *Trochammina pacifica* Cushman shows some increase. In the living assemblage the Rotaliina are typically eight times as abundant as the Textulariina. In the dead assemblage the reverse is true, and this may be due to postmortem solution of the calcareous tests. Two species do not tolerate polluted conditions and are more abundant on unpolluted shelf, namely *Nonionella miocenica stella* Cushman and Moyer and *Nonionella scapha basispinata* (Cushman and Moyer).

Deep basins

Because the sea floor between the mainland and the continental slope is so complex off southern California, Shepard and Emery (1941) termed it the continental borderland. In general, there are elongate basins separated by submerged or emergent banks. According to Emery (1960) the flat floors of the basins and troughs form 17 per cent of the total area of continental borderland. Slopes around these deep areas form approximately 63 per cent of the total area, while the shelves and bank tops comprise 20 per cent.

In each basin the properties of the water are similar to those in the overlying waters. The temperature is nearly uniform throughout each basin, but shallower basins are warmer than deeper ones. Salinity is close to 34 per mille. Dissolved oxygen is generally less than 1 cm³/l.

Bandy (1963) studied foraminiferids larger than 1 mm from the continental borderland. Although the study was not strictly of living or stained individuals, he stated that 'Protoplasmic material was noted in at least some of the specimens of each species . . .', and inferred that the species are indigenous. He recognized six faunal groups (Figure 56).

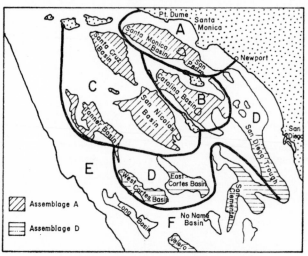

Figure 56 Map of the deep basins off southern California, showing the restricted distribution of parts of assemblages A and D (from Bandy, 1953).

Assemblage A Characteristic of San Pedro and Santa Monica basins. Bottom water temperature 5 °C, oxygen 0.2–0.3 cm³/l, salinity 34.3 per mille.

> *Bolivina spissa* Cushman
> *Bulimina affinis* d'Orbigny
> *Chilostomella ovoidea* Reuss
> *Globobulimina pacifica pacifica* Cushman
> *Globobulimina pyrula spinescens* (Brady)
> *Planulina ornata* (d'Orbigny)
> *Reophax agglutinans* Cushman
> *Robulus thalmanni* Hessland
> *Valvulineria inaequalis* (d'Orbigny)

Assemblage B In Santa Catalina basin the bottom water temperature is 4.2 °C, oxygen 0.4 cm³/l and salinity 34.2 per mille.

> *Dentalina communis* d'Orbigny
> *Hoeglundina elegans* (d'Orbigny)
> *Pyrgo murrhina* (Schwager)
> *Pyrgo ringens* (d'Orbigny)

Assemblage C   In the Santa Cruz and San Nicolas basins the bottom water temperature is 3.71–4.15 °C, oxygen 0.5–0.8 cm³/l and salinity 34.52 per mille.

> *Cibicides wuellerstorfi* (Schwager)
> *Cyclammina cancellata* Brady

Assemblage D   found in the southern and outer part of the continental borderland. Bottom temperature less than 3.5 °C, oxygen, 0.9–2.0 cm³/l, salinity 34.52–34.58 per mille.

> *Glandulina laevigata* d'Orbigny
> *Haplophragmoides* sp.
> *Martinottiella communis occidentalis* (Cushman)
> *Martinottiella communis pallida* (Cushman)

Assemblage E   found in the deepest basins (depth down to 2571 m). Temperature 2.5–2.6 °C, oxygen 1.3–2.0 cm³/l, salinity, 34.56–34.58 per mille.

> *Laticarinina pauperata* Parker and Jones

Assemblage F   found in the deepest basins (as for assemblage E).

> *Ammodiscus pacificus* Cushman and Valentine
> *Pyrgoella sphaera* (d'Orbigny)

Many of these species, particularly *Robulus thalmanni* Hessland, *Cyclammina cancellata* Brady and *Martinottiella* spp., show an increase in size with increasing depth. This agrees with the experimental results of Bradshaw (1961) who showed that there is an inverse relationship between size and temperature.

Two of the assemblages are mutually exclusive. The *Globobulimina pyrula spinescens–Globobulimina pacifica pacifica* components of assemblage A are confined to northern basins with sill depths shallower than 1200 m and temperatures higher than 3.85 °C. Assemblage D is found in the southern basins with sills deeper than 1200 m and temperatures less than 3.4 °C (Figure 56).

Resig (1958) took fourteen samples from the Santa Cruz Basin. On the west side of the basin at a depth of 140 m (edge of the basin), there are 275 living individuals per 100 g of sediment. At 730 m this is reduced to less than 20 per 100 g, and in the deepest parts (1500–1900 m) to 3–4 per 100 g. The importance of differentiating living and dead individuals is emphasized by Resig's statement that only rarely does the living distribution of a species correspond with that of the dead. Normally the living occurrences are close to the upper depth limit of the species and this suggests down-slope transport of empty tests.

The dominant species at depths less than 365 m are:

> *West side*
> *Cassidulina limbata* Cushman and Hughes
> *Cassidulina californica* Cushman and Hughes
> *Cassidulina tortuosa* Cushman and Hughes
> *Hoeglundina elegans* (d'Orbigny)
> *Cancris sagra* (d'Orbigny)

East side
   Cassidulina limbata Cushman and Hughes
   Cassidulina tortuosa Cushman and Hughes
   Rotorbinella lomaensis (Bandy)
   Robulus cf. *R. cultratus* Montfort
   Gyroidina io Resig
   Ammobaculites arenaria Natland

At the depth of the sill (1080 m) living species include:

   Globobulimina pacifica Cushman
   Pullenia salisburyi Stewart and Stewart
   Fursenkoina seminuda (Natland) (=*Virgulina seminuda* of Resig).
   Epistominella levicula Resig.
   Valvulineria inaequalis (d'Orbigny)
  Cassidulina delicata Cushman
   Cassidulina cushmani Stewart and Stewart
   Bolivina spissa Cushman
  Bolivinita minuta (Natland)
   Epistominella smithi (Stewart and Stewart)
   Uvigerina peregrina Cushman
  Fursenkoina nodosa (Stewart and Stewart) (=*Virgulina nodosa*
      of Resig).

* Indicates species confined to the depth of the sill.

Species found living deeper than 46 m listed with their depth ranges
(in metres) are

Cassidulina delicata Cushman	460–920, max. 730
Bolivina spissa Cushman	460–1300, max. 730
Cassidulina cushmani Stewart and Stewart	920–1300, max. 1100
Uvigerina peregrina Cushman	920–1650, max. 1460
Epistominella smithi (Stewart and Stewart)	460–1830, max. 1460
Epistominella levicula Resig	920–1860, max. 1860

Resig considers that those species living below the depth of the sill are
normally found in the bathyal zone (183–1830 m) in the open ocean. Lack
of zonation in the basin beneath sill depth is attributed to the lack of variation in environmental conditions. At the sill proper, there are fewer species
than at comparable depths elsewhere in the basin. This has been attributed
to the effects of water renewal over the sill.

One question that remains unanswered is whether the dead assemblages
of the basins are related to modern environmental conditions. Resig showed
that although there is a very rough division into two zones shallower and
deeper than 1080 m (sill depth) using either living or total assemblages, on a
finer scale there are big differences. Approximately half the living species
used to define depth zones are unimportant as dead individuals, while a

quarter of the dead species used to define depth zones have no living representatives. Also the depth zones of living species are more restricted than those of their dead representatives. Further studies of basinal assemblages are clearly needed.

BAJA CALIFORNIA

Todos Santos Bay

This large bay is bordered to the south and north by high ground, while to the east there are low sandy beaches and the marsh of Estero de Punta Banda. The floor of the bay dips seaward to a depth of approximately 80 m. Beyond this depth the slope steepens, and to the south-west leads to a 400 m deep, narrow channel between the Islas de Todos Santos and the Punta Banda mountains.

Surface water temperatures vary from a minimum of 14 °C to a maximum of 20 °C throughout the year. During June and July the temperature is warmest (16 °C) and upwelling occurs on the south side of Punta Banda. The cold upwelled water extends round the Islas de Todos Santos and a short distance into the bay (Walton, 1955). Surface salinities are 33.4–38.7 per mille.

Walton divided the sediments into 3 groups:

1. Coarse pebble and cobble deposits with epifaunal polyzoans and foraminiferids, shell debris and some fine detrital material;
2. the common sediment type of fine sands and silts;
3. muddy sediments found in the depth range 200–400 m.

Living foraminiferids were collected from 182 stations during 8 cruises in February, March, April, June, July, August, October and November, 1952. Walton (1955) recognized four bay assemblages:

Outer bay

> *Reophax gracilis* (Kiaer)
> *Uvigerina peregrina* Cushman vars.
> *Recurvoides* sp.
> *Chilostomella ovoidea* Reuss
> *Bolivina acuminata* Natland
> *Bolivina pacifica* Cushman and McCulloch
> *Bulimina denudata* Cushman and Parker
> *Globobulimina* spp.

This assemblage is restricted to the deeper bay and channel between the Islas de Todos Santos and Punta Banda (Figure 57). The depth is greater than 73 m (40 fms) and the sediment is group 3 muds and group 2 fine sands and silts.

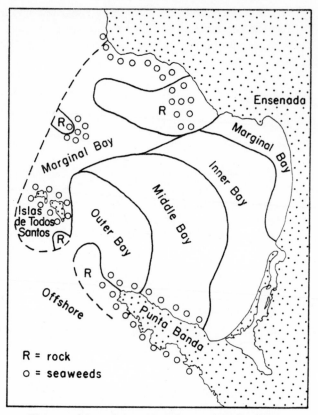

Figure 57 Map of the assemblages of Todos Santos Bay (based on Walton, 1955, and Benson, 1959).

Middle bay

    Goësella flintii Cushman
    Reophax curtus Cushman
    Proteonina sp.
    Ammotium planissimum (Cushman)
    Labrospira cf. *L. advena* (Cushman)
    Reophax scorpiurus Montfort

This assemblage lies in the depth range 18–73 m (10–40 fms) and mainly on the fine sands and silts of the group 2 sediments.

Inner bay

    Nonionella miocenica stella Cushman and Moyer
    Saccammina atlantica (Cushman) (=*Proteonina atlantica* of Walton)
    Nonionella basispinata (Cushman and Moyer)
    Discorbis spp.
    Labrospira cf. *L. columbiensis* (Cushman)
    Trochammina pacifica Cushman
    Eggerella advena (Cushman)
    Quinqueloculina sp.
    Elphidium translucens Natland
    Buliminella elegantissima (d'Orbigny)

These species occur throughout the bay but are most abundant in the inner bay. The sediment is mainly fine sand and silt of group 2 in the depth zone 0–27 m (0–15 fms).

Marginal bay

> *Trifarina angulosa* (Williamson) (=*Angulogerina angulosa* of Walton)
> *Cassidulina subglobosa* Brady
> *Cibicides fletcheri* Galloway and Wissler
> *Cibicidina nitidula* Brady
> *Elphidium tumidum* Natland
> '*Rotalia*' spp.
> *Bolivina striatella* Cushman
> *Textularia* cf. *T. schencki* Cushman and Valentine
> *Gaudryina* cf. *G. subglabrata* Cushman and McCulloch
> *Bolivina vaughani* Natland
> *Bifarina hancocki* Cushman and McCulloch
> *Planulina exorna* Phleger and Parker

This assemblage occurs in two areas, along the inner margin of the bay on the fine sands and silts of group 2 sediments, and between the north side of the bay and the Islas de Todos Santos in the area of group 1 sediments—the coarse pebble and cobble deposits.

In addition there is an offshore assemblage found between the Punta Banda—Islas de Todos Santos area at 360 m and seawards to at least 1100 m:

> *Cancris auricula* (Fichtel and Moll)
> *Valvulineria inequalis* (d'Orbigny)
> *Valvulineria araucana* (d'Orbigny)
> *Bolivina argentea* Cushman
> *Bolivina spissa* Cushman
> *Bolivina minuta* Natland
> *Cassidulina delicata* Cushman
> *Cassidulina laevigata* d'Orbigny
> *Bulimina exilis tenuata* (Cushman)
> *Fursenkoina bramletti* (Galloway and Morrey) (=*Virgulina bramletti* of Walton)
> *Fursenkoina seminuda* (Natland) (=*Virgulina seminuda* of Walton)
> *Reophax horrida* Cushman
> *Epistominella smithi* (Stewart and Stewart)
> *Loxostomum pseudobeyrichi* Cushman
> *Höglundina elegans* (d'Orbigny)

The standing crop size varies from 0 to 923 per 10 cm^2, but only 41 of the 182 samples have values greater than 100. Samples were collected seasonally along a traverse from Ensanada, in the north-east corner of the bay, across the bay, between the Islas de Todos Santos and Punta Banda and out to the sea in a south-westerly direction. Considerable variation in standing crop size is seen along the traverse at each season and the position of the maximum standing crop varies with depth at each season, the maxima occurring at depths of 18–700 m. Walton averaged the data for each depth zone and found three maxima: more than 200 individuals per 10 cm² at 55–73 m, 142 per 10 cm² at 90–180 m and 80 per 10 cm² at 640–730 m. However, it is doubtful whether these values are meaningful as they obscure the very large variation found in each depth zone at a single season and they are greatly influenced by the big seasonal variations. The main feature of significance is that August is the season of maximum abundance.

A triangular plot of the forty-one samples with more than 100 living individuals shows the Rotaliina and Textulariina to be the important suborders and the Miliolina to be poorly represented (Figure 58). This may be

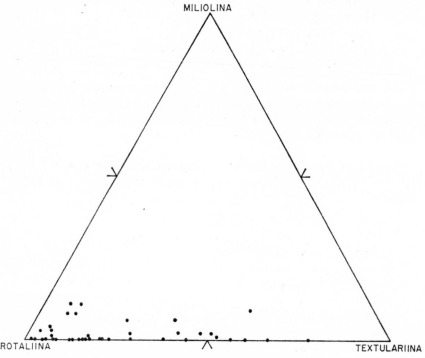

Figure 58 Triangular plot of Todos Santos Bay samples.

partly attributable to the slightly low salinities (33 per mille). The diversity shows a total range of α values from 1.5 to 9.5 (Figure 59) but there appears to be some depth control. Three overlapping ranges are evident: 0–36 m, α = 1.5 to α = 5.5; 36–180 m, α = 2.5 to α = 9.5; and >180 m, α = 2.5 to α = 6. In practice the highest α values are found in the 36–180 m zone, although low values are found throughout all three depth zones. A similarity matrix for the 41 valid samples shows values generally less than 50 per cent. This may indicate that the samples are not representative or it may truly reflect a patchy distribution of species within the assemblages.

Four depth associations were recognized by Walton with boundaries at 55 m (30 fms), 91 m (50 fms), 180 m (100 fms) and 640 m (350 fms). Two features were thought to be conspicuous:

1. 'the relatively greater number of species at shoaler depths, and
2. the wider distribution of the deep species'.

More than 50 per cent of the species are confined to or are most abundant in the 0–90 m depth zone.

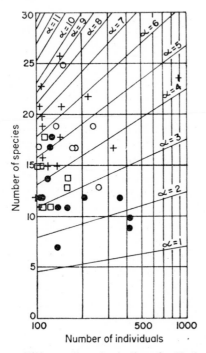

Figure 59 Fisher α diversity indices for Todos Santos Bay. ● = 0–35 m, + = 36–71 m, ○ = 72–180 m, □ = >180 m.

There is a considerable difference between the living and total assemblages. Apart from differences of relative abundance for individual species, more than ten species have no living representatives in the bay area, and in the offshore region there are eleven such species. Some were present in abundance. In addition there are fossil individuals reworked from submarine and land outcrops of Tertiary strata.

Kaesler (1969) used Walton's data, together with those of Benson (1959) on the ostracods, to carry out a quantitative computer study. When negative matches are taken into consideration (e.g. the absence of a normally common species from two stations can be interpreted as similarity of unsuitability for that species) the dendrogram resulting from the use of simple matching coefficients closely agrees with Walton's biotope map and his separation of inner, middle, outer and marginal bay assemblages. However, using the Jaccard coefficient dendrogram, the assemblages at the 0.20 level of correlation disagree markedly with those proposed by Walton.

GULF OF CALIFORNIA

The Gulf is about 900 km long and averages about 150 km across. It is a region of faulting and subsidence, and the submarine topography is characterized by fault scarps, ridges and basins. In the axial region depths vary from 2900–3600 m near the mouth to less than 180 m in the north. A continental shelf is developed only on the eastern side (Rusnak, Fisher and Shepard, 1964).

The land surrounding the Gulf is arid. The mountains of Baja California reduce the influence of the Pacific Ocean on the climate. Thus conditions are continental rather than oceanic. A number of rivers discharge into the Gulf, but evaporation exceeds fresh-water input so that salinities are higher than those of the adjacent Pacific. Over much of the Gulf values are around 35 per mille throughout the year, but during the summer months values of 36 per mille occur in the northernmost part. During the rainy season, coastal salinities are reduced to less than 34 per mille. Low-salinity water is thought to flow in along the east coast and high-salinity water to flow out along the west coast (between 50 and 100 m depth). Surface temperatures in the northern part are 15–30 °C while in the south the range is 21–30 °C. Local wind-induced upwelling causes variations in surface temperature. A thermocline is developed in the upper 50–100 m of the water column. At a depth of 200 m temperatures are around 12 °C, at 500 m, 8 °C, and at 800 m, 5.5 °C (data from Roden, 1964).

Phleger (1964a) studied seventy-six samples from the area. Only twenty-seven contained more than 100 individuals. These together with data from Phleger (1965c) have been used for diversity and triangular plots.

The standing crop varies from 2 to 2948 per 10 cm². The data for different depth zones are summarized in Table 16. Low values are found at depths

Table 16 Standing crop data for a unit area of 10 cm² (data from Phleger, 1964a, 1965c)

Depth (m)	Minimum	Maximum	Number of stations	Average	
>1800	2	62	25	26.5	
900–1800	4	44	13	25.5	Phleger (1964a)
180–900	56	702	4	238.5	
0–180	19	632	31	237.7	
180–594	13	722	8	166.0	Phleger (1965c)
0–180	6	2948	47	282.5	

greater than 900 m. At shallower depths the maximum values rise to 2948 per 10 cm² and the average values for the depth zones 900–180 m are 238.5 and 166 and for the zone 180–0 m are 237.7 and 282.5 per 10 cm². Particularly high values are found off Rio Fuerte (492 and 702 per 10 cm²), in the northern Gulf (370–632 per 10 cm²) and at unspecified localities off Nyarit, Sinaloa and Sonora, Mexico.

List of Plates

Plate 1 Hyposaline marshes

1. *Ammotium salsum* (Cushman and Brönnimann), ×130 (San Antonio Bay, Texas)

2, 3. *Miliammina fusca* (Brady), 2. ×120, 3. ×140 (Christchurch Harbour, England)

4, 5, 6. *Arenoparrella mexicana* (Kornfeld), 4. ×170, 5. ×170, 6. ×170 (Mississippi Sound)

7, 8. *Trochammina inflata* (Montagu), 7. ×105, 8. ×105 (Christchurch Harbour, England)

9, 10, 11. *Jadammina macrescens* (Brady), 9. ×100, 10. ×150, 11. ×200 (Christchurch Harbour, England)

Plate 2 Hyposaline lagoons

1, 2. *Ammonia beccarii* (Linné), 1. ×115, 2. ×115 (Bristol Channel, England)
3. *Protelphidium anglicum* Murray, ×200 (Christchurch Harbour, England)
4, 5. *Elphidium articulatum* (d'Orbigny), 4. ×175, 5. ×175 (Christchurch Harbour, England)

Plate 3 Hypersaline lagoons

1. *Peneroplis planatus* (Fichtel and Moll), ×50 (Abu Dhabi)
2. *Elphidium* cf. *E. discoidale* (d'Orbigny), ×110 (Abu Dhabi)
3, 4, 5. *Quinqueloculina* type B of Murray (1966b), 3. ×120, 4. ×105, 5, ×120 (Abu Dhabi)
6. *Trioculina* type C of Murray (1966b), ×110. Dried protoplasm partly fills the aperture (Abu Dhabi)
7, 8. *Rosalina adhaerens* Murray, 7. ×265, 8. ×270 (Abu Dhabi)
9, 10. *Archaias angulatus* (Fichtel and Moll), 9. young individual ×76, 2. adult ×39 (Jamaica)

Plate 4 Larger foraminiferids: tropical forms

1. *Marginopora vertebralis* Blainville, ×20 (Seychelles)
2. *Borelis pulchra* (d'Orbigny), ×160 (Jamaica)
3. *Baculogypsina sphaerulata* (Parker and Jones), ×50 (Gilbert Islands)
4. *Operculina ammonoides* (Gronovius), ×50 (Zanzibar)
5, 6. *Calcarina spengleri* (Gmelin), 5. ×55, 6. ×55 (Marshall Islands)

Plate 5 Shelf seas: Atlantic seaboard of North America

1, 2, 5. *Elphidium subarcticum* Cushman, 1. ×335, 2. ×130, 5. ×130
3, 4. *Elphidium clavatum* Cushman, 3. ×175, 4. ×175
6. *Globobulimina auriculata* Bailey, ×130
7, 8. *Eggerella advena* (Cushman), 6. ×230, 7. aperture ×1500

Plate 6 Shelf seas: Atlantic seaboard of North America

1, 2. *Cribrostomoides crassimargo* (Norman), 1. ×100, 2. ×75
3. *Fursenkoina fusiformis* (Williamson), ×260
4, 5, 6. *Saccammina atlantica* (Cushman), 4. ×130, 5. ×185, 6. ×190
7, 8. *Buccella frigida* (Cushman), 7. ×160, 8. ×170

Plate 7 Shelf seas: Atlantic seaboard of North America

1, 2. *Cibicides concentricus* (Cushman), 1. ×170, 2. ×170
3, 4. *Cibicides pseudoungerianus* (Cushman), 3. ×175, 4. ×175
5, 6. *Stetsonia minuta* Parker, 5. ×300, 6. ×350
7, 8. *Florilus grateloupi* (d'Orbigny), 7. ×140, 8. ×140

Plate 8 Shelf seas: European seaboard

1, 2. *Ammonia beccarii* (Linné), 1. ×96, 2. ×90
3, 4, 5. *Nonionella turgida* (Williamson), 3. ×150, 4. ×165, 5. ×165
6. *Cassidulina carinata* Silvestri, ×170
7. *Fursenkoina fusiformis* (Williamson), ×190
8. *Cassidulina obtusa* Williamson, ×330
9. *Eggerella scabra* (Williamson), ×70
10, 11. *Trochammina globigeriniformis* (Parker and Jones) var. *pygmaea* Höglund,
10. ×250, 11. ×250

Plate 9 Shelf seas: European seaboard

1. *Bulimina* cf. *B. alazanensis* Cushman, ×100
2. *Hyalinea balthica* (Schröter), ×80
3. *Bulimina gibba/elongata* Fornasini/d'Orbigny, ×105
4. *Cibicides lobatulus* (Walker and Jacob), ×70
5. *Cribrostomoides jeffreysii* (Williamson), ×100
6. *Textularia sagittula* Defrance, ×80
7, 8. *Asterigerinata mamilla* (Williamson), 7. ×105, 8. ×105
9. *Brizalina spathulata* (Williamson), ×90
10. *Melonis pompilioides* (Fichtel and Moll), ×80
11. *Trifarina angulosa* (Williamson), ×100

Plate 10 Shelf seas: Pacific seaboard of North America

1, 2. *Nonionella basispinata* (Cushman and Moyer), 1. ×80, 2. ×85
3. *Nonionella stella* Cushman and Moyer, ×150
4. *Bolivina acuminata* Natland, ×140
5, 6. *Trochammina pacifica* Cushman, 5. ×150, 6. ×150
7. *Bolivina subargentea* Uchio, ×90
8. *Bolivina spissa* Cushman, megalospheric, ×115
9. *Chilostomella ovoidea* Reuss, ×100
10. *Buliminella elegantissima* (d'Orbigny), ×190
11. *Bolivina pacifica* Cushman and McCulloch, ×120
12. *Alveolophragmium colombiense* (Cushman), ×100

Plate 11 Damaged foraminiferid tests

1, 2. *Archaias angulatus* (Fichtel and Moll) showing the effects of abrasion causing (*a*) removal of the outer wall and (*b*) rounding, 1. ×90, 2. ×65 (Jamaica)

3. *Archaias compressus* (d'Orbigny) showing borings made by algae, ×70 (Jamaica)

4. *Quinqueloculina seminulum* (Linné) showing borings made by a predator, ×105 (English Channel)

5. *Gaudryina rudis* Wright etched in 5 per cent EDTA. The calcareous cement has partially dissolved. Solution of this kind may take place during fossilization, ×60 (English Channel)

6. *Protelphidium anglicum* Murray etched in 5 per cent EDTA causing solution of the calcite layers of the radial lamellar wall. Solution of this kind is commonly observed in fossil forms, ×2000 (Christchurch Harbour, England)

Plate 12 Attachment

1. *Protelphidium anglicum* Murray with a turbellarian egg case attached during life, ×140 (Christchurch Harbour, England)

2. *Globulina caribaea* d'Orbigny cemented to a quartz sand grain, ×160 (Cape Hatteras)

3. *Acervulina* sp. attached to *Julienella*, ×2000 (Ghana)

4. *Rosalina* sp. clinging to a sand grain, ×280 (Jamaica)

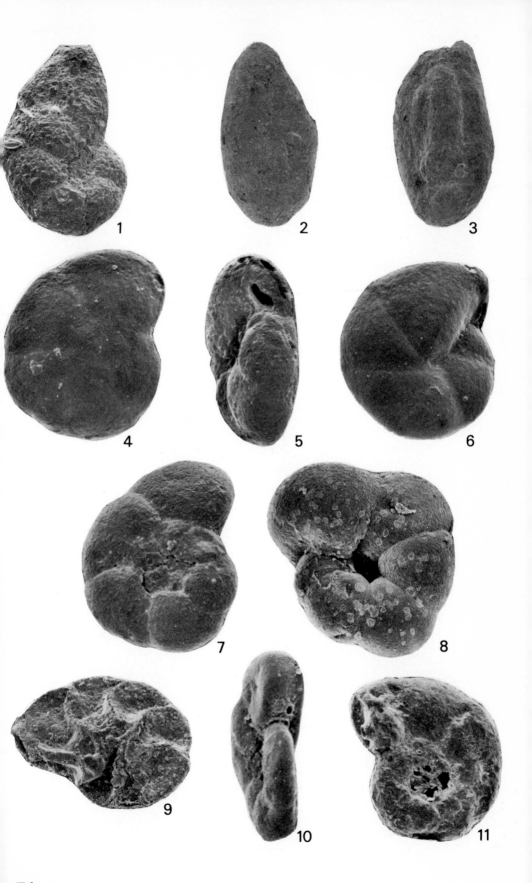

Plate 1

Plate 2

Plate 3

Plate 4

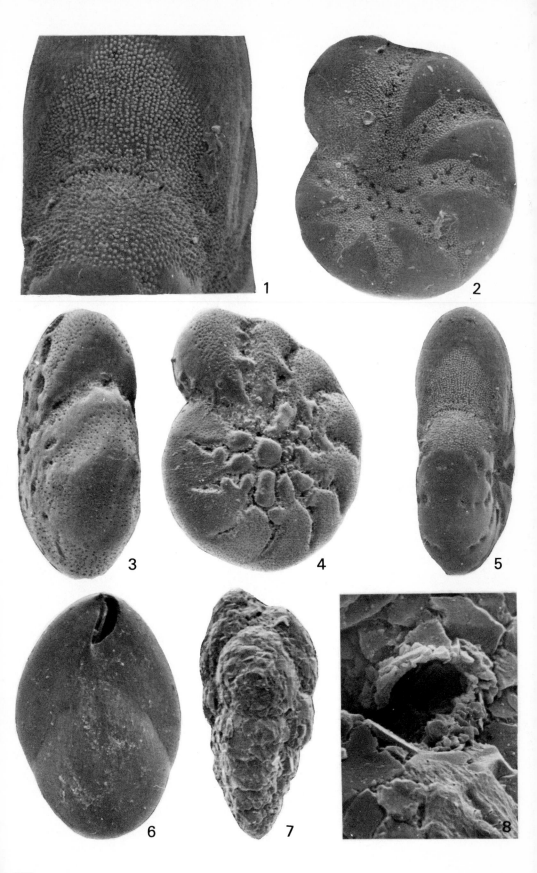

Plate 5

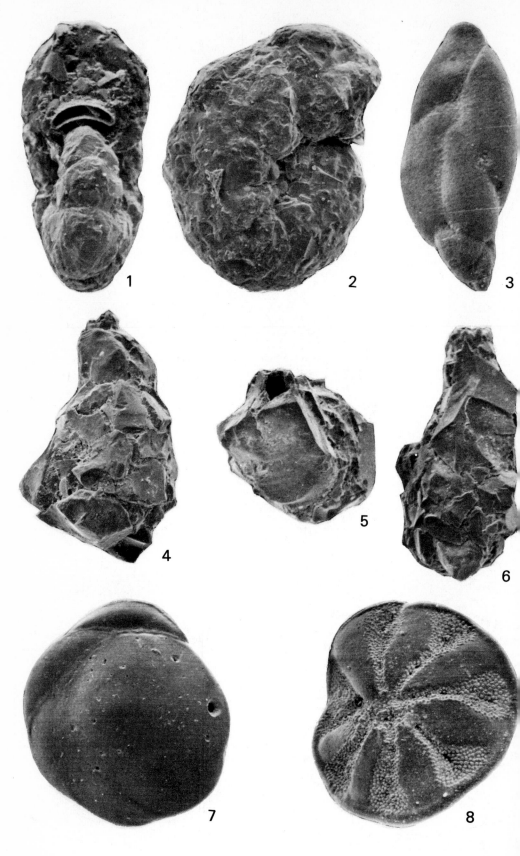

Plate 6

Plate 7

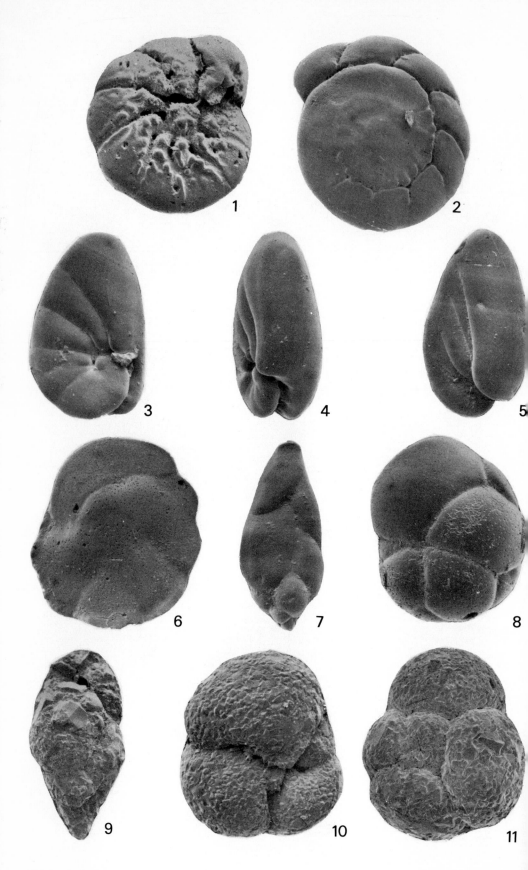

Plate 8

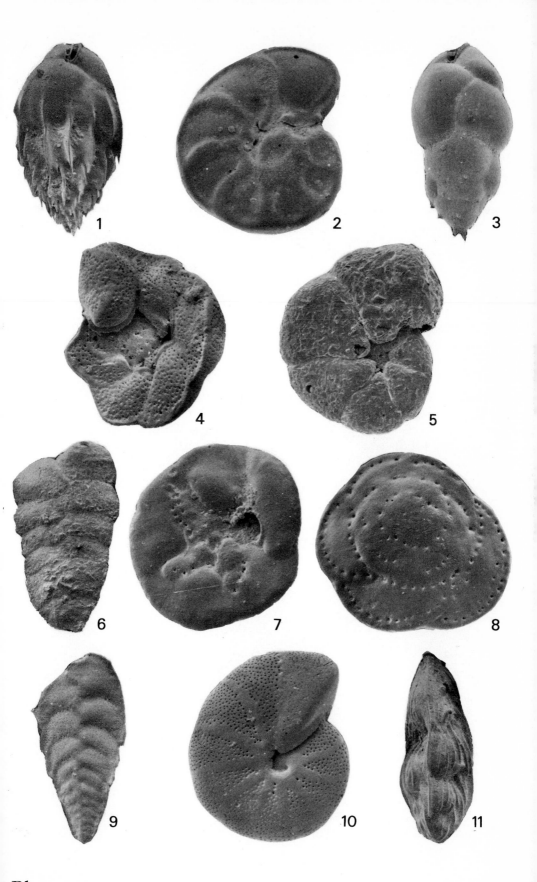

Plate 9

Plate 10

Plate 11

Plate 12

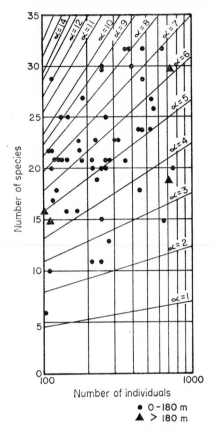

Figure 60 Fisher α diversity values for the Gulf of California.

The Fisher α diversity values range from 1.5 to 12 for the depth range 0–180 m, to 3.5 to 6.5 for depths greater than 180 m (Figure 60). On the triangular plot, samples from the nearshore shallow-water zone (0–33 m) have an unusual composition with a mixture of all three suborders. Other deeper water shelf samples are poor in Miliolacea, have 0–51 per cent Textulariina and 49–100 per cent Rotaliina. Samples from deeper than 180 m have high Rotaliina, low Textulariina and no Miliolina (Figure 61).

Phleger (1964a) recognized a series of depth zones with limits at 27–32 m, 55–64 m, 73–82 m, 130 m, 165 m, 365 m, 730–820 m, 1100–1460 m, 2400 m, and 2750 m. Only the major limits are discussed here.

5.5 to 27–32 m zone The assemblages are very varied and no species can be selected as typical of the majority of samples, although Miliolina are common. The following species reach their deepest limit at 27–32 m:

Ammonia beccarii (Linné) in abundance
Buliminella elegantissima (d'Orbigny) in abundance
Triloculina inflata d'Orbigny
Quinqueloculina laevigata d'Orbigny

Elphidium translucens Natland
Reussella pacifica Cushman and McCulloch
Quinqueloculina lamarckiana d'Orbigny
Nonionella basispinata (Cushman and Moyer) in abundance
Trochammina kellettae Thalmann in abundance

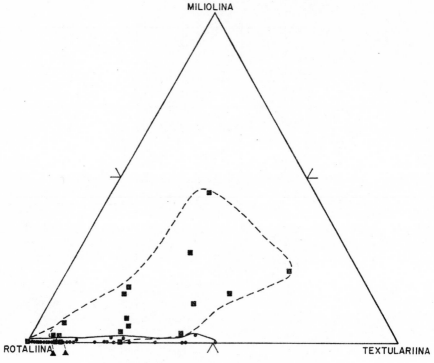

Figure 61 Triangular plot of Gulf of California
samples (data from Phleger, 1964a). ■ = 0–33 m,
● = 33–180 m, ▲ = > 180 m.

32–73 m zone The species of this zone are present also at depths less than
32 m, but there they are associated with the characteristic shallow zone
species. The species named below reach their deepest limit at around 73 m:

Buccella tenerrima (Bandy)
Textularia schencki Cushman and Valentine group
Ammoscalaria pseudospiralis (Williamson)
Rotorbinella spp.
Fursenkoina pontoni (Cushman)

Many other species that extend into deeper water are present. In particular *Nonionella stella* Cushman and Moyer is abundant.

73–130 m zone Species which reach their deeper limit in this zone are:

Bulimina denudata Cushman and Parker
Bolivina vaughani Natland

The most abundant species is *Nonionella stella* Cushman and Moyer.

130–730 m zone   With the exception of *Nonionella stella* Cushman and Moyer all the above named species are absent. New arrivals in this zone include:

> *Globobulimina pacifica* Cushman
> *Bolivina pacifica* Cushman and McCulloch
> *Fursenkoina seminuda* (Natland) at depths greater than 360 m

730–1100 m zone   The species of the previous zone continue with the addition of *Fursenkoina spinosa* (Heron-Allen and Earland). *Nonionella stella* Cushman and Moyer dies out at the base of the zone.

1100–2750 m   The dominant species are:

> *Bolivina pacifica* Cushman and McCulloch
> *Fursenkoina spinosa* (Heron-Allen and Earland)
> *Eponides leviculus* (Resig)

It is of interest that Parker (1964) recognized depth zones based on macro-invertebrates at 11–26 m, 27–65 m, 66–120 m, 121–730 m, 731–1799 m and 1800–4122 m. There is considerable agreement between the foraminiferids and the macro-invertebrates in this respect.

GEOGRAPHIC DISTRIBUTION

There is no very obvious faunal change between the Gulf of California and the southern California region. Some of the differences which exist are probably due to differences of nomenclature. Phleger (1964a) is the only author to have commented on the species showing a limit of distribution. He noted that *Cancris panamaensis* Natland, *Epistominella obesa* Bandy and Arnal and *Eponides antillarum* (d'Orbigny) are southern species which extend up into the Gulf of California. However, as this is an eastern oceanic border, faunal boundaries are likely to be diffuse and only a detailed study of species distribution will locate them.

FERTILITY OF THE SEA FLOOR

The San Diego area stands apart from Todos Santos Bay and the Gulf of California with respect to the size of the standing crop. The very high values recorded there contrast strongly with the average values of the latter areas. This is probably a direct result of the upwelling which occurs off San Diego. The upwelling leads to increased phytoplankton production, which in turn favours other organisms, including the benthic foraminiferids.

NOTE ON CENTRAL AMERICAN PACIFIC COAST

Off El Salvador, duplicate samples from a series of depths, 20–3200 m, were studied by Smith (1963, 1964). On the continental shelf the sediments are mainly very fine sands; on the slope they are mainly silt and clay.

Oceanographic data collected at the same time as the samples revealed differences of temperature and dissolved oxygen in the bottom water. Salinity was greater than 33 per mille and less than 35 per mille. The thermocline is developed between 60 and 80 m in December and rises to 25–50 m during June, but the temperature variation is small from one season to another.

The standing crop varies from 7 to 243 per 10 cm^2 but in only 4 of the 18 samples is it greater than 100 per 10 cm^2 (Smith, 1964). The Rotaliina dominate the assemblages, Miliolina are rare and only one sample, from a depth of 885 m on sand, shows an appreciable percentage of Textulariina (20 per cent).

Smith contends that abundant dead species are represented by some living individuals, and therefore the six depth zones recognized are considered to be valid. In view of the general sparse occurrence of living forms it seems best to ignore these zones. For the same reason the comparative analysis of depth zones along the North and Central American coast is not reproduced here.

6. Shelf Seas: Atlantic Seaboard of North America

The southern tip of Florida lies at latitude 25 °N while the easterly Cape of Newfoundland lies at latitude 47 °N. Throughout this latitudinal spread of 22°, the coastline may be divided into three geographic units: Florida–Cape Hatteras, Cape Hatteras–Cape Cod, Cape Cod–Newfoundland (Figure 62).

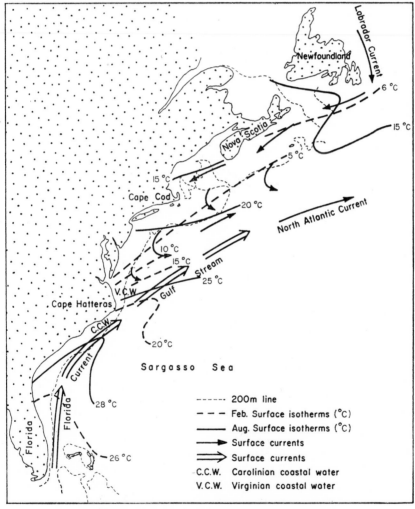

Figure 62 The eastern seaboard of North America with physical parameters of the water (based on data from Sverdrup, Johnson and Fleming, 1942).

Water circulation in the North Atlantic is dominated by the north-easterly flowing warm Gulf stream and the south-flowing cold Labrador Current. In the straits of Florida, the Florida Current is initiated partly by the difference in sea level between the Gulf of Mexico and the adjacent Atlantic coast, and partly by the circulation of the atmosphere (Sverdrup,

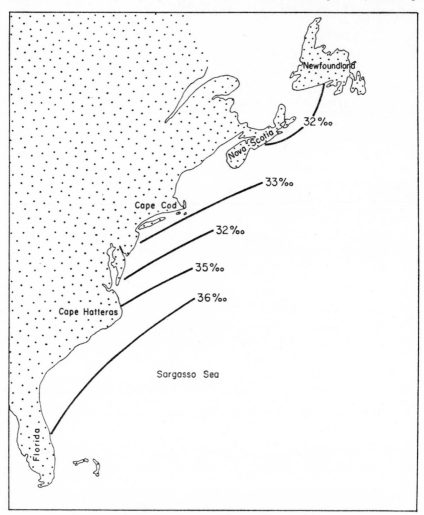

Figure 63 Salinity conditions during the summer
(based on Sverdrup, Johnson and Fleming, 1942).

Johnson and Fleming, 1942). The current closely follows the continental slope as far as Cape Hatteras. It is increased in volume by the addition of the Antilles Current and Sargasso Sea water. At Cape Hatteras, the Florida Current diverges from the coast and becomes known as the Gulf Stream.

The Labrador Current is a south-moving cold water mass which flows round Newfoundland and thence down the coast past Nova Scotia. Inter-action between these two major currents, together with local physiographic differences, leads to the development of well defined nearshore water masses such as the Carolinian coastal water, Virginian coastal water and the cold water pool off Long Island. These are discussed with the foraminiferid data.

In Figure 62, the surface water isotherms have been plotted for August and January. These are not necessarily meaningful for benthic organisms unless the water is isothermal. However, the temperature range experienced at the surface can give a crude indication of subsurface conditions.

Off the Florida coast, the difference between February and August is 26 to 28 or 29 °C (2 to 3 °C). At Cape Hatteras it is 20 to 25 °C (5 °C). Off New York it is 5 to 20 °C (15 °C). In the Gulf of Maine it is 1 to 15 °C (14 °C). For shallow, nearshore benthic organisms, conditions south of Cape Hatteras are clearly more stable than to the north.

Examination of the surface salinity data shows that north of Cape Hatteras salinity decreases from 35.0 per mille to 32.5 per mille at Cape Cod (Figure 63). In the Gulf of Maine values of 32 per mille prevail.

The regions from which living foraminiferids have been described are outlined in Figure 64.

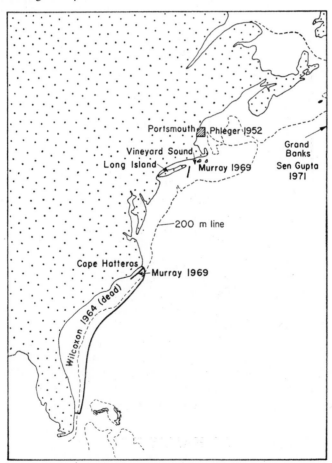

Figure 64 Areas of study of foraminiferids on the eastern seaboard of North America.

GRAND BANKS, NEWFOUNDLAND

The area studied by Sen Gupta and McMullen (1969) and Sen Gupta (1971) is a shallow region, 50–80 m depth, called the Tail of the Grand Banks. Temperatures recorded at the time of sampling varied from −0.9 to 8.3 °C. The salinity is 33 per mille. The substrate is mainly sandy, with areas of muddy sand and gravel.

The diversity of the twenty-one samples with a standing crop of more than 100 individuals, is low, with α values from 1 to 4.5 (Figure 65). The

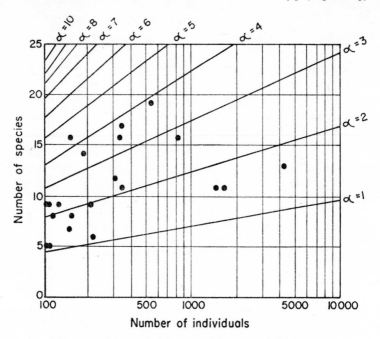

Figure 65 Fisher α diversity values for the Tail of the Grand Banks (data from Sen Gupta, 1971).

triangular plot shows dominance of the Rotaliina (Figure 66). The dominant species are *Islandiella islandica* (Nørvang), *Elphidium clavatum* Cushman, *Nonionellina labradorica* (Dawson), *Globobulimina auriculata* (Bailey), and *Eggerella advena* (Cushman). Locally abundant species include *Cibicides lobatulus* (Walker and Jacob), *Islandiella teretis* (Tappan) and *Fursenkoina loeblichi* (Feyling Hansen) (=*Virgulina loeblichi* of Sen Gupta). Only 43 of the 88 benthic species had living representatives. The standing crop is 1–4156 per 10 cm² with average values of 100–200 per 10 cm². The mixing of warm and cold water masses in this area probably accounts for the higher standing crop values in the western part of the area.

PORTSMOUTH, NEW HAMPSHIRE

Phleger (1952) examined the living foraminiferid assemblages in 213 samples. The nearshore area is sandy in the south but passes into gravel and rock to the north. Offshore glacial till forms topographic highs, and in the deeper basinal areas (>50 m) mud and muddy sand predominates. The water is isolated from the Gulf Stream by topographic barriers. Isothermal conditions prevail during the winter months, with mean values of 2 °C nearshore and 4 °C offshore during February. A thermocline develops by May and the highest temperatures are reached during August, with a surface

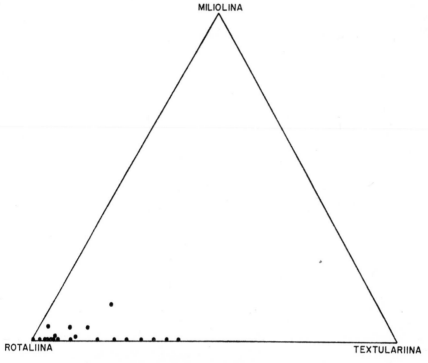

Figure 66 Triangular plot of the Tail of the Grand Banks (data from Sen Gupta, 1971).

average of 15.5–17 °C. The thermocline extends down to 40 or 50 m and the bottom water temperature is about 5.5 °C. Salinities are slightly low throughout the year, being 32.0–32.5 per mille in the winter, 30 per mille nearshore to 32 per mille offshore in April, and 29–31.5 per mille in the summer and autumn.

The samples were collected from depths of 9–180 m. Living foraminiferids were identified by using Millon's reagent to stain the protoplasm red. Unfortunately this involves use of nitric acid which caused the destruction of calcareous shells, leaving only the agglutinated species. For this reason the results obtained by this method are not directly comparable with those obtained with rose Bengal.

The twenty-six living species encountered represent 40 per cent of the total species found. All the common species are represented by living individuals, and Phleger concluded that the geographical distribution of the total assemblage is closely similar to that of the living assemblage. The largest standing crops occur in three areas in which mud-sand sediment is present. The range of standing crop size is 1–111 per 10 cm² but the majority of values are less than 50 per 10 cm². The diversity of the three samples with more than 100 individuals is $\alpha = 1$ to $\alpha = 2$ (based only on agglutinated species).

Phleger recognized two principal facies, namely on sand and on mud or mud-sand bottoms. From the data on total assemblages the following species are found to be dominant:

Sand facies
    Cassidulina algida Cushman
    Cibicides lobatulus (Walker and Jacob)
    Elphidium articulatum (d'Orbigny)
    Elphidium clavatum Cushman
    Elphidium subarcticum Cushman

Mud facies
    Haplophragmoides bradyi (Robertson)
    Haplophragmoides glomeratum (Brady)
    Cribrostomoides crassimargo (Norman) (=*Labrospira crassimargo* of Phleger)
    Cribrostomoides jeffreysii (Williamson) (=*Labrospira jeffreysii* of Phleger)
    Recurvoides turbinatus (Brady)
    Reophax curtus Cushman
    Reophax scottii Chaster
    Spiroplectammina biformis (Parker and Jones)
    Textularia torquata Parker

Many other species are present in low abundance in each facies, and in addition there are local variations in assemblages related to the presence of rivers, shoals, etc.

The sand facies occurs mainly at depths of less than 50 m and therefore it is subject to large temperature variations during the course of a year. By contrast, the mud facies is found at depths greater than 50 m and therefore deeper than the thermocline. Temperature conditions are less variable (3–6 °C) and always cold. The assemblage here is dominated entirely by agglutinated species.

VINEYARD SOUND

This channel lies between the Elizabeth Islands and Martha's Vineyard Island. To the south-west is the Atlantic, and to the north-east is Nantucket Sound. Depths in the central sound reach a maximum of 35 m. The powerful tidal currents reach surface speeds of 2–3 knots. The bottom currents produce sand waves up to 10 m high. The sediments are sands and gravels derived from glacial deposits. The mean monthly salinity is 32 per mille. The temperature at the time of sample collection was 16.0–16.6 °C and conditions were more or less isothermal. However, a thermocline is present from June to September. During the winter, the bottom water is warmer than that at the surface.

Murray (1969) studied a traverse of stations along the length of the Sound. The living assemblage is dominated by *Eggerella advena* (Cushman) with *Rosalina* spp. and *Trochammina ochracea* (Williamson) somewhat less abundant. The standing crop at four stations is 7, 63, 66 and 147 per 10 cm² with a biomass of 0.007, 0.052, 0.147 and 0.346 mm³ per 10 cm² respectively. The diversity indices range from α = 3 to α = 6 (Figure 67). A plot

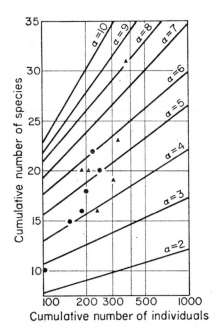

Figure 67 Diversity plot of Vineyard Sound samples.
● living ▲ dead.

of the ratios of the suborders shows a distinct field with high Rotaliina and Textulariina and low Miliolina (Figure 68). Similarity indices between adjacent samples is 48.6–82.0 per cent.

It is believed that the following species live clinging to pebbles, shell debris or some other suitable substrate:

Cyclogyra involvens (Reuss)
Patellina corrugata Williamson
Rosalina spp.
Spirillina vivipara Ehrenberg
Trochammina ochracea (Williamson)
Wiesnerella auriculata (Egger)

These species form 20.8–51.2 per cent of the living assemblage. The presence of species with a clinging mode of life seems to be characteristic of high-energy environments subject to powerful currents.

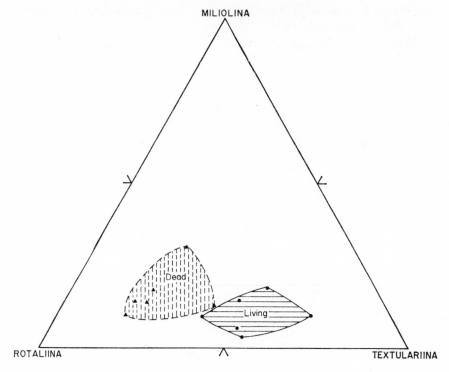

Figure 68 Triangular plot of Vineyard Sound
samples. ● living ▲ dead.

THE SHELF OFF LONG ISLAND

Murray (1969) sampled a line of stations from near Block Island out across
the shelf. This is an area with unusual hydrographic conditions. At the time
of sampling (September 1966) a thermocline was present and bottom water
temperatures ranged from 11.4 °C near the shore to 8.0 °C away from the
shore (Figure 69). Ketchum and Corwin (1964) have described the annual
changes:

> 'During the winter season, temperatures are vertically homogeneous from
> about December to April or May; the minimum temperature generally
> occurs in March. From May to July, the surface waters warm rapidly,
> and then more slowly, reaching maximum values in mid-August or
> September. During July and August, the thermocline gradually deepens
> as greater volumes of surface water are warmed. The warming of the
> deep waters is associated with this gradual deepening of the thermocline
> and with a decrease in intensity of the gradients toward the end of sum-
> mer. Between September and November, the density structure becomes
> sufficiently diffuse that complete vertical mixing again occurs. The
> admixture of the warmer surface waters produces the maximum tem-
> perature in the deep water at this time, a month or more after the

maximum surface water temperature. The water then remains nearly vertically homogeneous throughout the winter, cooling gradually. During this period the deep water is often slightly warmer than the surface water.'

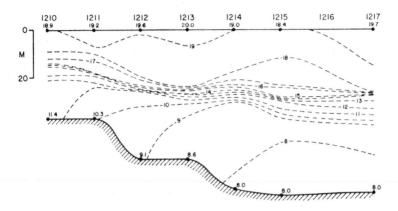

Figure 69 Temperature relationships along the shelf traverse off Long Island, September 1966.

Thus a cold-water pool is formed on the shelf in the winter and persists throughout the summer as first demonstrated by Bigelow (1933).

Salinities are in the range 32 to 35 per mille with the lower values near the land. Tidal currents are small. The sediments range from sands to muddy sands and are partly relict (Shephard and Cohee, 1936; McMaster and Garrison, 1966).

Murray found twelve dominant species whose occurrence is shown in Figure 70. *Cribrostomoides crassimargo* (Norman) and *Elphidium clavatum* Cushman both show a fairly uniform occurrence throughout the traverse. *Eggerella advena* (Cushman), *Reophax scottii* Chaster, *Saccammina atlantica* (Cushman), *Quinqueloculina* spp. and *Miliolinella* spp. show a high absolute abundance at the shoreward end of the traverse and become progressively less abundant seaward. *Fursenkoina fusiformis* (Williamson), *Globobulimina auricula* (Bailey) and *Nonionella* cf. *N. turgida* (Williamson) are common at the seaward end of the traverse.

Similarity indices between adjacent samples range from 34.3 per cent to 73.6 per cent. The lower value suggests a faunal break. It coincides with the incoming of *G. auricula* and *Nonionella* cf. *N. turgida*, the decrease in abundance of several species, and a change in the substrate from sand to muddy sand.

Diversity is $\alpha = 4$ to $\alpha = 5$ seaward of the faunal break, and $\alpha = 5$ to $\alpha = 6$ landward (Figure 71). At the faunal break where the two assemblages overlap, diversity reaches a maximum at $\alpha = 11$. The standing crop varies from 11 to 95 per 10 cm² and biomass 0.003 to 0.634 mm³ per 10 cm².

The plot of the ratio of the three suborders (Figure 72) shows two fields: stations 1210–4 have a predominance of Textulariina while stations 1215–17 have a predominance of Rotaliina.

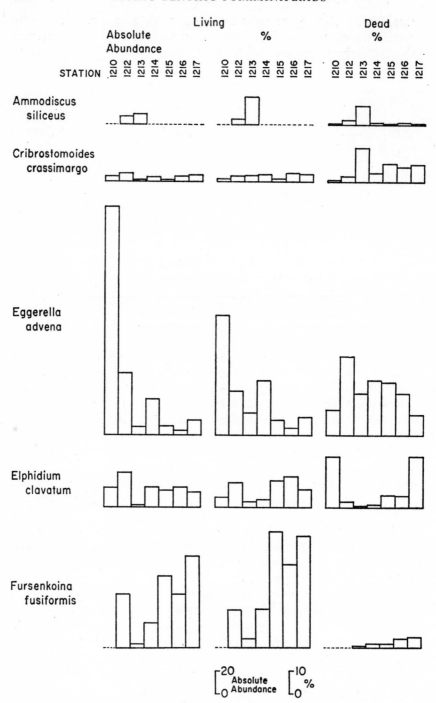

Figure 70 (Part 1) Histogram of the dominant species in the shelf traverse off Long Island.

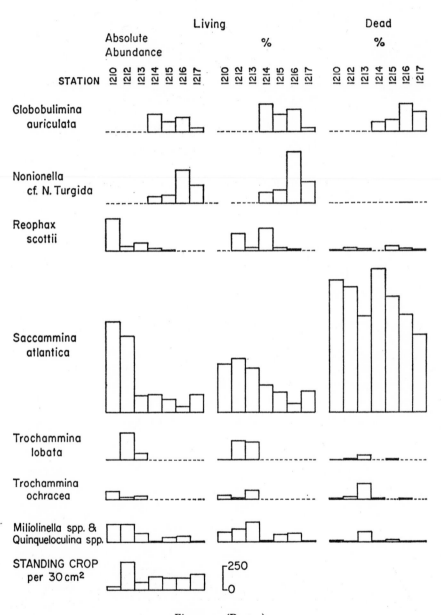

Figure 70 (Part 2)

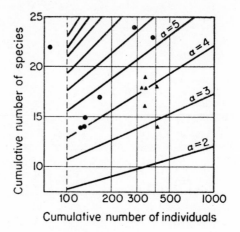

Figure 71 Diversity plot of the Long Island traverse.
● living ▲ dead.

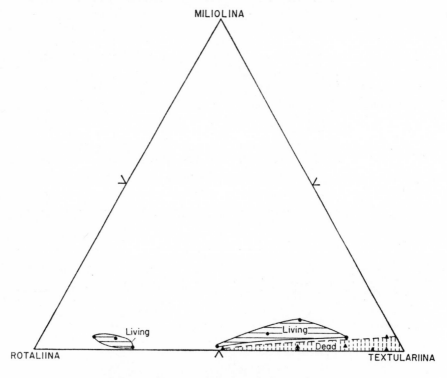

Figure 72 Triangular plot of the Long Island
traverse. ● living ▲ dead.

The faunal break between stations 1213 and 1214 seems clearly defined on a number of criteria and these are summarized in Table 17.

Table 17 Summary of the assemblages on the shelf off Long Island to show the faunal break

Landward
Eggerella advena
Elphidium clavatum
Fursenkoina fusiformis
Reophax scottii
Saccammina atlantica
Trochammina lobata
Miliolinella spp.
Quinqueloculina spp.

α ranges from 5 to 6, similarity 48.6–53.6 per cent, standing crop 11–95 per 10 cm², depth <55 m, temp. 8.6–11.4 °C, sand.

Faunal break
Sta. 1213

α = 11, similarity 34.3 per cent.

Seaward
Elphidium clavatum
Fursenkoina fusiformis
Globobulimina auriculata
Nonionella cf. N. turgida
Saccammina atlantica

α ranges from 4 to 5, similarity 53.3–73.6 per cent, standing crop 43–55 per 10 cm², depth >66 m, temp. 8.0 °C, muddy sand.

CAPE HATTERAS

A series of stations was sampled between Cape Hatteras, Diamond Shoals, across Raleigh Bay to Cape Lookout Shoals to the south (Murray, 1969). The Diamond Shoals mark the position of a strong horizontal temperature gradient between cooler waters to the north and warmer waters to the immediate south (Figure 73).

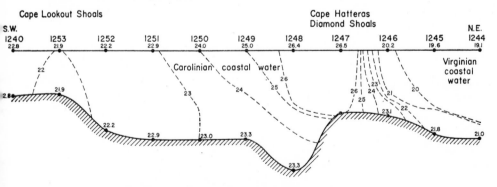

Figure 73 Temperature relationships along the traverse from north of Cape Hatteras to south of Cape Lookout, October 1966.

To the south of Cape Hatteras is the Florida Current, which flows north along the edge of the continental shelf. At Cape Hatteras it becomes the Gulf Stream and diverges away from the shelf. Between the Florida Current and the shore is the Carolinian coastal water, while between the Gulf Stream and the shore is the Virginian coastal water. This is cooler and less saline (31–34 per mille) than the Carolinian water (35 per mille). The sediments are mainly sand and may be partly relict.

Very few species occur at every station, and many do not show any very marked trends. *Ammonia beccarii* (Linné) and *Buliminella elegantissima* (d'Orbigny) both develop local peaks of abundance and are unimportant elsewhere. In the Virginian coastal water are:

Brizalina sp.
Cibicides concentricus (Cushman)
Eggerella advena (Cushman)
Rosalina spp.
Stetsonia minuta Parker

These species are present in the Carolinian water also, together with:

Cibicides pseudoungerianus (Cushman)
Florilus grateloupi (d'Orbigny)
Miliolinella circularis (Bornemann)
Quinqueloculina spp.
Trifarina angulosa (Williamson)

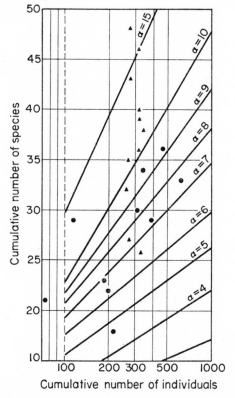

Figure 74 Diversity plot of the Cape Hatteras traverse. ● living ▲ dead.

The similarity indices are very different between adjacent stations, but this is to be expected in a region of faunal change. The diversity of indices range from α = 5 to α = 13 (Figure 74). Standing crop varies from 5 to 112 per 10 cm². The samples occupy a small field on the plot of the ratio of suborders (Figure 75).

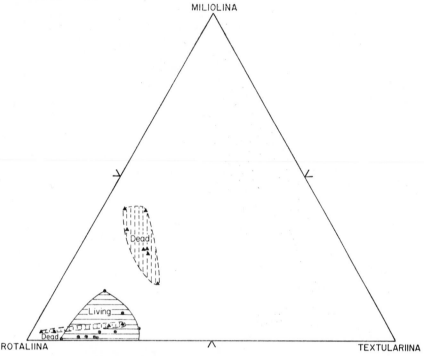

Figure 75 Triangular plot of the Cape Hatteras traverse. ● living ▲ dead.

GEOGRAPHIC DISTRIBUTION

Even with the limited amount of data available, it is possible to comment on some of the geographic distributions observed in the shelf assemblages. We are not particularly concerned with the cosmopolitan species which are found along the entire seaboard, but rather with these species which have a restricted distribution. Five groups are listed below and their boundaries are shown diagrammatically in Figure 76.

1. Species found north of Cape Cod and in the Arctic (based on Phleger (1952), and Leslie (1965) and Sen Gupta (1971):

Ammotium cassis (Parker)
Astrononion stellatum Cushman and Edwards
Cassidulina algida Cushman
Cassidulina islandica Nørvang var. *minuta* Nørvang
Cribrostomoides jeffreysii (Williamson)
Nonion labradoricum (Dawson)
Reophax arctica Brady
Trochammina quadriloba Högland

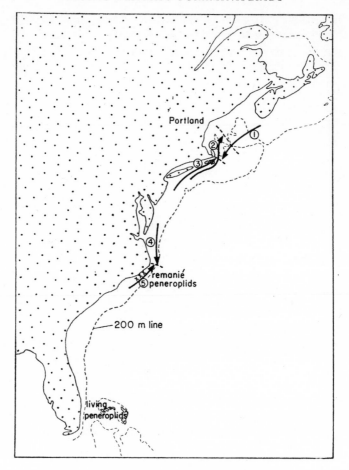

Figure 76 The position of faunal boundaries of the
five groups of species discussed in the text.

2. Species reaching their northern limit near the Portsmouth region (based
on Phleger (1952) and Leslie (1965)):

> *Ammodiscus minutissimus* Cushman and McCulloch
> *Bolivina pseudoplicata* Heron-Allen and Earland
> *Elphidium advenum* (Cushman) var. *margaritaceum* Cushman
> *Elphidium excavatum* (Terquem)
> *Eponides umbonatus* (Reuss)
> *Haplophragmoides bradyi* (Robertson)
> *Labrospira* cf. *nitida* (Goës)
> *Miliammina fusca* (Brady)
> *Trochammina advena* Cushman
> *Trochammina lobata* Cushman
> *Jadammina macrescens* (Brady)

3. Species reaching their northern limit at Cape Cod (based on a comparison of Phleger (1952) and Murray (1969)):

Ammonia beccarii (Linné)
Cassidulina crassa d'Orbigny
Cibicides pseudoungerianus (Cushman)
Elphidium macellum (Fichtel and Moll)
Epistominella exigua (Brady)
Poroeponides lateralis (Terquem)

4. Species which appear to reach their southern limit near Cape Hatteras (based on data from Murray (1969) and Parker (1948)):

Buccella frigida (Cushman)
Cribrostomoides crassimargo (Norman)
Elphidium clavatum Cushman
Globobulimina auriculata (Bailey)
Reophax curtus Cushman
Reophax scottii Chaster
Saccammina atlantica (Cushman)
possibly *Eggerella advena* (Cushman)

5. Species reaching their northern limit at Cape Hatteras (based on a comparison of Murray (1969) and Wilcoxon (1964)):

Asterigerina carinata d'Orbigny
Elphidium poeyanum (d'Orbigny)
Florilus grateloupi (d'Orbigny)
Reussella spinulosa (Reuss)
Sagrina pulchella d'Orbigny

In addition there are dead representatives of *Peneroplis, Vertebralina, Articulina* and *Quinqueloculina* in the vicinity of Cape Hatteras. The present northern limit of these forms must lie somewhere to the south as they are known to live in the Florida-Bahama region.

Group 1 comprises cold-water species that live in slightly subsaline waters (salinity 32 per mille). Groups 2 and 3 are temperate species which again tolerate slightly subsaline conditions (down to 32 per mille). Group 4 comprises arctic and temperate species that reach their southern limit near Cape Hatteras. Group 5 includes subtropical southern species that will not tolerate the reduced salinities and temperatures north of Cape Hatteras. Thus a general correlation exists between the overall environmental conditions and the geographic distribution of the shelf foraminiferids.

POSTMORTEM CHANGES

In the Portsmouth region Phleger (1952, p. 356) concluded that there is good correlation between the distribution of the living and dead assemblages. However, south of Cape Cod Murray (1969) noted a number of differences.

In Vineyard Sound, the relative proportions of some of the species are different. *Eggerella advena* (Cushman) is less abundant dead while *Elphidium clavatum* Cushman and *E. subarcticum* Cushman are more abundant. At each sampling station, there are big differences between the living and dead assemblages (similarity indices of 35–47 per cent with one exception of 63 per cent). Diversity is higher with α values from 3 to 8 (Figure 67). The ratio of the three suborders shows separate fields with the living assemblages dominated by the Textulariina and the dead by the Rotaliina (Figure 68). It seems probable that a large part of these differences can be accounted for by the winnowing away of the smaller dead tests by the powerful tidal currents.

The shelf off Long Island also shows the loss of smaller species such as *Fursenkoina fusiformis* (Williamson), *Nonionella* cf. *N. turgida* (Williamson) and *Reophax scottii* Chaster from the dead assemblages. *Eggerella advena* (Cushman) and *Saccammina atlantica* (Cushman) are more abundant dead. *Cribrostomoides crassimargo* (Norman) is more abundant seawards. The biomodal distribution of dead *Elphidium clavatum* Cushman is quite unlike that of the living (Figure 70). Comparison of the living and dead assemblages at each station shows similarity indices of 28–53 per cent. Thus the assemblages are considerably different. The diversity values are $\alpha = 2$ to $\alpha = 5$ as compared with $\alpha = 4$ to $\alpha = 6$ for the living, except one value of $\alpha = 11$ at the faunal break). This reduction in diversity suggests removal of tests (e.g. of smaller species) without the introduction of tests derived from elsewhere.

Large differences between the living and dead assemblages are apparent at Cape Hatteras. Again, the smaller species are less abundant dead. *Cibicides concentricus* (Cushman), *Quinqueloculina* spp. and *Textularia agglutinans* d'Orbigny are more abundant dead. Two species which are scarcely represented by living individuals are also abundant: *Elphidium clavatum* Cushman and *E. poeyanum* (d'Orbigny). In addition, *Peneroplis carinatus* d'Orbigny is present and is believed to be remanié. The complete lack of similarity between the living and dead assemblages at each station is shown by the low similarity indices of 17–48 per cent. Diversity is markedly higher, with α values from 6.5 to 17 (cf. α values from 5 to 13 for the living, Figure 74). The triangular plot is different too (Figure 75). The likely causes of these postmortem differences are transport of dead tests, fluctuations in the position of the currents and water masses leading to variations in living assemblages with time, and the presence of remanié foraminiferids from Pleistocene sediments.

7. Shelf Seas: Gulf of Mexico

NORTHERN GULF

The northern Gulf of Mexico was one of the earliest-studied areas of continental shelf, and several large papers by Bandy (1956), Parker (1954), Phleger (1951, 1960b) and Walton (1964) have supplied much data on the distribution of dead and total assemblages. Unfortunately, the amount of information on living assemblages is small. Since there are known to be relict late Pleistocene and Holocene forms in the modern dead assemblages (Ludwick and Walton, 1957, Walton, 1964a) the interpretation of dead or total assemblages must be treated with caution.

The surface waters of the Gulf are slightly hyposaline near the coasts, but bottom salinities are generally normal (36 per mille). Surface temperatures reach 30 °C in the summer and fall to 19 °C in the northern coastal regions in winter.

Phleger (1951) studied samples from over a wide area of the north-west Gulf of Mexico. At the time of sampling the rose Bengal staining method for differentiating living foraminiferids had not been developed. Phleger used the Biuret test. The number of living individuals recorded was very small (see Phleger, 1951, Tables 22–29) so the data cannot be re-interpreted by the methods used in this book.

Walton (1964a, pp. 195–201) summarized the occurrence of living forms in a review paper. Of the 543 living assemblages studied, the majority were from shallow water. The largest standing crop, averaging 180 per 10 cm^2, occurs in the depth range 11–18 m. Locally values reach 1000 per 10 cm^2. In deeper water the abundance decreases to approximately 10 individuals per 10 cm^2 at 360 m. The dominant inner shelf species are *Nonionella opima* Cushman, *Nouria polymorphinoides* Heron-Allen and *Hanzawaia strattoni* (Applin). On the outer shelf species of *Bolivina*, *Bulimina* and *Uvigerina* dominate, while at depths greater than 360 m, *Chilostomella* and *Globobulimina* are important. In shallow water miliolids increase in abundance towards Florida, and Walton interprets this as a 'West Indian' influence.

In the north-east Gulf of Mexico, at a depth of 70–100 m on the shelf, is a zone of 'drowned' reefs. Ludwick and Walton (1957) took forty-one samples from this tract south of Alabama. The bottom temperature was 16–18 °C and salinity 37 per mille. The standing crop ranges from 10 to 149 per 10 cm^2 and averages 63 per 10 cm^2. Only five samples contained more than 100 living foraminiferids, and these give α values of 10 to 22, which are very high. These assemblages are dominated by the Rotaliina, while the Miliolina are very rare. *Bolivina* and *Nonionella* are the dominant genera. Approximately one quarter of the species are not represented by living individuals. The forms found only in the dead condition include shallow water species and relict West Indian varieties such as *Asterigerina carinata* d'Orbigny, *Nodobaculariella cassis* (d'Orbigny), peneroplids and *Amphistegina*.

Shifflet (1961) studied a similar bank, the Heald Bank, off Galveston, Texas. She listed fifty species which were found living on the sediment. Of these *Elphidium discoidale* (d'Orbigny), *E. poeyanum* (d'Orbigny) and *Hanzawaia strattoni* (Applin) were most abundant.

CENTRAL TEXAS SHELF

Along this coast a series of hyposaline lagoons are separated from the waters of the Gulf of Mexico by an almost continuous barrier of islands. The shelf sediments are poorly sorted silty clays or sandy silty clays. Salinities are around 36 per mille except in the nearshore zone where values may fall to 30 per mille in regions of land runoff.

Phleger (1956) described the assemblages from three main traverses across the shelf and took widely spaced samples along four other traverses. The greatest depth of water sampled was 110 m.

Only thirty samples had a standing crop greater than 100 per 10 cm². The average value for stations in each traverse is 21 per 10 cm² in the south, 44, 61 and 103 in the north-east. However, the latter value is artificially high

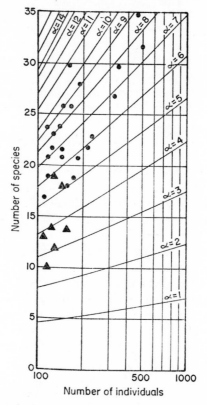

Figure 77 Diversity plot of Central Texas Shelf (data from Phleger, 1956). ▲ = <21 m, ● = >21 m.

as one station yielded 938 per 10 cm². Thus much of the area has low fertility.

The α diversity indices range from 2.5 to 6 at nearshore stations at depths of less than 21 m and 5–11 in the depth range 21–110 m (Figure 77). On the triangular plot (Figure 78) the samples fall into a field dominated by the

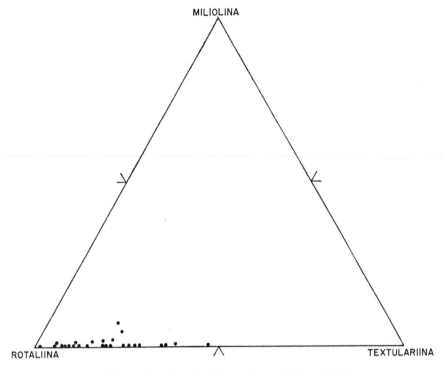

MILIOLINA

ROTALIINA

TEXTULARIINA

Figure 78 Triangular plot of Central Texas Shelf (data from Phleger, 1956).

Rotaliina and to a lesser extent the Textulariina. Few samples contain Miliolina, and only one station, from a depth of 16 m, has more than 5 per cent.

Phleger recognized depth boundaries at 20–30 m, 50–70 m, and 100 m (approximately) based on total assemblages. The dominant living species are listed below together with their main depth range:

Bigenerina irregularis Phleger and Parker	10–80 m
Ammonia beccarii (Linné)	
='Rotalia' beccari vars. A and B of Phleger	8–35 m
Nonionella atlantica Cushman	10–72 m
Nonionella opima Cushman	8–80 m
Nouria polymorphinoides Heron-Allen and Earland	15–40 m
Fursenkoina pontoni (Cushman)	
=*Virgulina pontoni* of Phleger	15–80 m
Hanzawaia strattoni (Applin)	22–78 m

Cancris oblonga (Williamson)	45–75 m
Bolivina striatula spinata Cushman	55–75 m
Bolivina fragilis Phleger and Parker	52–100 m

From this it may be seen that the boundaries are barely recognizable from the living assemblages because most of the dominant species extend over a wide depth range. Buzas (1967) applied canonical analysis to the results and confirmed that Phleger's depth boundaries are valid only for the total assemblage. He concluded that the living assemblages in the north-east shallow, nearshore area were different from the remainder. Further computer studies by Mello and Buzas (1968) using Q- and R-mode cluster analysis showed good agreement for the depth boundaries of the living (12 m and 77 m) and total (10 m and 77 m) assemblages.

8. Shelf Seas: Mediterranean

The Mediterranean is a land-locked sea in communication with the Atlantic Ocean only through the relatively shallow Straits of Gibraltar. Evaporation exceeds runoff, so that at depth there is higher density water which flows over the Straits of Gibraltar sill while Atlantic water flows in at the surface (Sverdrup, Johnson and Fleming, 1942). Mixing of the Atlantic water with the Mediterranean water soon causes it to lose its Atlantic character. The surface salinity is 37 per mille except in areas where fresh water is added in large amounts (inner portion of Adriatic and Aegean Seas). The surface isotherms are shown on Figure 79. Seasonal temperature variation always exceeds 9 °C and reaches up to 14 °C off the French Riviera.

Sverdrup, Johnson and Fleming describe four water layers. The surface layers extending down to 100–200 m is characterized by vertical convection especially during the winter. The intermediate water extending down to about 600 m is characterized by a salinity maximum at 300–400 m. Temperature and salinity decrease at the base of the layer and throughout the underlying transition layer to a depth of 1500–2000 m. Below these depths is the deep water with temperatures close to 13 °C and a salinity greater than 38 per mille.

The most important study of living foraminiferids in the Mediterranean is that by Blanc-Vernet (1969). She studied areas around Marseille, Corsica, the Aegean Sea and Crete, Peleponese and Rhodes. Samples were taken also from other parts of the northern Mediterranean. Le Calvez and le Calvez (1958) studied the miliolids of Villefranche Bay.

NEARSHORE SEDIMENT AND FLORA (down to about 40 m)

Living individuals have been found in the superficial sediment, on the leaves of *Cymodocea* and on tufts of *Posidonia* in the Marseille region, and in particular at Baie du Brusq and Port d'Alon.

On the sediment *Ammonia beccarii* (Linné) is present throughout the year, although it is rare in the winter. *Eggerella scabra* (Williamson) is also present although rare in January and February. *Elphidium macellum* (Fichtel and Moll) occurs from March to October, particularly as the spiny variety *aculeatum*. In July and August *Elphidium excavatum* (Terquem) and *E. advena* (Cushman) are present in small numbers. Blanc-Vernet states (1969, p. 29) that *Elphidium* has not been found living on *Cymodocea* and its habitat seems to be exclusively on the sediment. Various miliolids are also encountered in small numbers throughout the year.

Thus the coastal sediment away from areas of vegetation is not a favourable habitat for most benthic foraminiferids.

On vegetation, miliolids are abundant and the presence of peneroplids

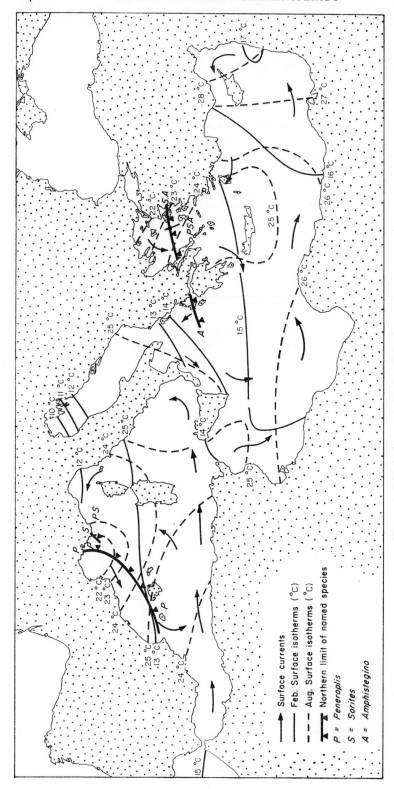

Figure 79 Physical oceanography (based on Sverdrup, Johnson and Fleming, 1942) and distribution of the larger foraminiferids in the Mediterranean.

such as *Peneroplis planatus* (Fichtel and Moll), *P. pertusus* (Forskål) and *Spirolina* is important. They are sparsely present on the sediment in winter, but at the beginning of the summer they undergo a rapid development and reach their maximum in September–October. At the same time *Vertebralina striata* d'Orbigny becomes abundant. It is believed that the increased water temperature and the calm conditions of the summer are the cause of increase of these species. At Port d'Alon, *Sorites variabilis* Lacroix is also present in the very warm conditions there. This species is found on the Côte d'Azur in association with *Penicillus mediterranensis* and *Caulerpa olivieri*. Further to the east in the still warmer waters around Crete, the southern Aegean Sea and the Dodecanese, these species are abundant and the presence of *Amphistegina madagascariensis* d'Orbigny makes the fauna almost subtropical.

A number of adherent species are also found on the *Cymodocea*, notably *Planorbulina mediterranensis* d'Orbigny, *P. acervalis* Brady, *Cibicides lobatulus* (Walker and Jacob) and various discorbids. Many other than these are of small size. In the autumn, the *Cymodocea* dies off and most of the foraminiferids are contributed to the sediment.

In the eastern Mediterranean two stations on pure *Cymodocea* (i.e. no *Halophila*) at Baie du Milieu (north of Crete), depth 0.5–1 m, and south of Crete, depth 7 m, showed the same microfauna. The leaves bore numerous small discorbids, *Peneroplis*, *Sorites*, *Nubecularia* and *Vertebralina*. There is little difference with the coast of Provence except for their more general occurrence.

In the eastern Mediterranean, along the south and east coasts of Peloponnisos, the coasts of Crete, the central part of the Aegean Sea up to 38° latitude, the Dodecanese and Rhodes, the phanerogam *Halophila stipulacea* occurs in association with *Cymodocea*. Here the warmer water enables an almost subtropical assemblage of *Amphistegina madagascariensis* d'Orbigny and *Sorites variabilis* Lacroix to colonize these plants. On the east coast of Crete, at a depth of 10 m, the flora comprises *Halophila*, *Caulerpa* and Rhodophycean algae. The common living forms are:

	Halophila	*Caulerpa*	Rhodophyceans
Amphistegina	VA	P	VA
Peneroplis	A	A	A
Miliolidae	P	P	VA
Planorbulina	A	A	—
Cibicides	A	A	—
Discorbidae	A	A	—
Nubecularia	A	A	A
Iridia	A	A	A
Elphidium	A	A	A

VA=very abundant, A=abundant, P=present

At the same locality, the assemblage on the sediment is rich in *Amphistegina* (although less so than the vegetation), *Rotalia granulata* (di Napoli Alliata) and *Ammonia beccarii* (Linné). Some *Ammonia* individuals approach *A.* cf. *gaimardii* (d'Orbigny).

Along the Provence coast, thickets of *Posidonia* provide a habitat for a diverse microfauna. Common species include:

12.3% { *Nubecularia lucifuga* Defrance
Vertebralina striata d'Orbigny
Cyclogyra involvens (Reuss) (=*Cornuspira involvens* of Blanc-Vernet)

21.5% { *Quinqueloculina* 20 spp.
Massilina 2 spp.
Triloculina 12 spp.
Sigmoilina 2 spp.
Spiroloculina 4 spp.

9.4% { *Peneroplis planatus* (Fichtel and Moll)
Peneroplis pertusus (Forskål)

0.9% { *Astrononion sidebottomi* Cushman and Edwards
Nonion asterizans (Fichtel and Moll)

23.4% { Discorbidae—various

16.8% { *Cibicides* spp.

Asterigerinata mamilla (Williamson)
Elphidium spp.

6.3% { *Planorbulina mediterranensis* d'Orbigny
Acervulina inhaerens Schultze

Ammonia beccarii (Linné)
Miniacina miniacea (Pallas)

(Percentages relate to data for Port d'Alon, after Blanc-Vernet, 1969, Figure 5A).

Posidonia seems to be an important environment for the reproduction of many foraminiferids. The same seasonal occurence of peneroplids is observed as on the *Cymodocea*.

In the warmer waters around Peloponnisos, Crete and the Aegean Sea, the fauna of *Posidonia* and *Halophila* is *Amphistegina, Sorites* and *Ammonia* cf. *A. gaimardii* (d'Orbigny). *Articulina, Elphidium, Parrellina* and *Glabratella* are common.

In the northern part of the Aegean Sea, *Amphistegina* and *Sorites* disappear and *Peneroplis* is less abundant. However, *Ammonia* cf. *A. gaimardii* (d'Orbigny) becomes very abundant and conical forms (*A. convexa* Collins) appear.

A great many algae are present in the nearshore zone of the northern Mediterranean, and during the summer they commonly bear an epiphytic microfauna. Along rocky coastlines subject to wave attack, *Cystoseira* supports few living foraminiferids—rare *Cibicides* and miliolids. *Miniacina miniacea* (Pallas) finds this habitat favourable. At Port d'Alon, rocks are coated with a thin layer of *Jania* and foraminiferids, many of the latter being empty tests. Miliolids are present almost to the exclusion of all other forms. Elsewhere *Halopteris* grows on sand and yields a rich microfauna. The seasonal increase in *Peneroplis* is most conspicuous, 4.5 per cent abundance in July increasing to 35.5 per cent in September.

AMPHIOXUS SANDS

Coarse sands and fine gravels with *Amphioxus* are found in situations where there are currents. The depth varies from a few metres to 100 m in channels between islands. Blanc-Vernet (1969) studied three stations in the Riou archipelago off Provence. The depths varied from 7–18 m. The sediment here is composed of shell debris of molluscs, echinoids and polyzoans together with fragments of *Lithothamnium*.

Neither dead nor living foraminiferids are very abundant. The standing crop is 30–40 per 10 cm². Miliolids are irregularly distributed. Discorbids are present. *Cibicides lobatulus* (Walker and Jacob), *C. refulgens* (Montfort) and *Planorbulina mediterranensis* d'Orbigny are frequently found fixed to stones and shells. They may originally have been attached to leaves of *Posidonia* and transported to the *Amphioxus* sands, where they survive for a while. *Textularia* 'agglutinans' d'Orbigny and *T. sagittula* Defrance are present. Blanc-Vernet concludes that the living assemblage cannot really be distinguished from that of the vegetation. Further, postmortem changes favour the large species which thus accumulate in the sediment. In the eastern Mediterranean, *Amphistegina mediterranensis* d'Orbigny can make up to 80 per cent of the sediment. Also sands and gravel of this kind receive much detrital biogenic material from adjacent areas of living animals.

DETRITAL SANDS

Beneath the vegetation zone, the shelf bears either detrital sands or muds.

The detrital sands extend from about 40 m down to about 100 m. Although dead, empty tests are common, the living assemblage is poor. Typical forms are:

> *Cibicides lobatulus* (Walker and Jacob)
> *Cibicides refulgens* (Montfort)
> *Cibicidella variabilis* (d'Orbigny)
> *Planorbulina mediterranensis* d'Orbigny
> *Quinqueloculina* spp.
> *Triloculina* spp.
> Discorbidae
> *Textularia* spp.
> *Guadryina rudis* Wright
> *Guadryina pseudoturris* (Cushman)

The presence of mud-loving forms such as *Cassidulina*, *Bulimina*, *Bolivina* and particularly *Nonion parkeri* le Calvez is dependent on the mud content of the sand.

The dead assemblage includes many remanié species not found living in this environment at present.

On the outer part of the continental shelf there is a belt of shelly sand. This is divisible into two parts, an upper zone at a depth of 100–150 m and a lower zone which extends from 150 m to the top of the bathyal muds. Living foraminiferids are rare here, and Blanc-Vernet interprets the dead assemblages as being of Quaternary age.

TERRIGENOUS MUDS ON THE SHELF

Blanc-Vernet (1969) studied samples from off the Rhône delta, in the Baie de Marseille and in the eastern Mediterranean, particularly around Crete, Rhodes and the Aegean Sea.

Typical terrigenous muds extend from approximately 30 or 40 m to 100 m depth. At greater depths they pass into bathyal muds. The fauna is very consistent throughout the area of mud distribution and the living assemblage is basically the same as the dead.

Reophax scottii Chaster forms up to 40 per cent of the living assemblage although locally, as in the Baie de Marseille, it is absent. Species confined to this muddy environment include *Nonionella turgida* (Williamson) and *N. miocenica* Cushman var. *stella* Cushman and Moyer. They are present at all stations at 45–99 m depth in the Golfe de Fos by the Rhône delta and in the Baie de Marseille. Their maximum development is at 75–80 m, and below 100 m they disappear completely.

Species forming more than 5 per cent of the assemblage are:

> *Reophax scottii* Chaster
> *Eggerella scabra* (Williamson)
> *Textularia* spp.
> *Bolivina* spp.
> *Bulimina* spp.
> *Nonionella turgida* (Williamson)
> *Nonionella miocenica* Cushman var. *stella* Cushman and Moyer
> *Nonion parkeri* le Calvez
> *Valvulineria bradyana* (Fornasini)

Other forms which characterize the shelf muds are *Hyalinea balthica* (Schröter) and *Cassidulina* spp.

Where mixed mud and sand occurs on the shelf the living assemblage is poor.

DEEP-WATER MUDS

The depth range considered here is 300–4200 m. Conditions are believed to be very uniform and the assemblages of foraminiferids show little variation. Although Blanc-Vernet (1969) stained for living individuals, she concluded that the total assemblage is a reliable guide as there are few remanié individuals. Planktonic species make up to 40–100 per cent of the total assemblage.

Off Provence, at a depth of 370–2400 m, there is a very diverse microfauna. Various species of *Cassidulina* form up to 10 per cent of the assemblage. *Bolivina* spp. and *Bulimina* spp., together with the rare species *Uvigerina auberiana* d'Orbigny, *U. mediterranea* Hofker and *Rectuvigerina* sp. make up to 20–30 per cent. *Lenticulina* spp., *Marginulina* spp., *Vaginulina* spp., *Nodosaria* spp. and *Dentalina* spp. form 10 per cent. The miliolids are represented by *Pyrgo* spp. and *Biloculinella* spp. and to a lesser extent by

Sigmoilina. Hyalinea balthica (Schröter) is present throughout although its abundance is related to differences of depth. The Textulariina are abundantly represented by *Saccammina sphaerica* Sars, *Psammosphaera fusca* Schultze, *Marsipella cylindrica* Brady, *Astrorhiza arenaria* Brady, *Rhabdammina* spp., *Hyperammina friabilis* Brady, *Bathysiphon filiformis* Sars, *Adercotryma glomerata* (Brady), *Cribrostomoides scitulum* (Brady), *Siphotextularia concava* (Karrer), *Bigenerina* spp. and *Cyclammina cancellata* Brady.

Samples from the Golfe de Gênes, off Corsica, and the eastern Mediterranean have essentially the same deep-water fauna.

Some influence of depth on the microfauna was noticed. *Hyalinea balthica* (Schröter) and *Lagenodosaria scalaris* Batsch increase in abundance to a depth of 500 m and then they become less frequent and form only 1–2 per cent of the microfauna at the deepest stations. Other forms such as *Pyrgo* and the agglutinated species, increase in abundance with increasing depth. Blanc-Vernet gives the following optimal depth ranges:

> *Uvigerina mediterranea:* abundant particularly at 400–1000 m
> *Hyalinea balthica:* abundant particularly at 200–600 m
> *Lagenodosaria scalaris:* abundant particularly at 400–700 m

Depth zonation

Zone I 0–40 m approximately

Ammonia beccarii (Linné)
Rotalia granulata (di Napoli Alliata)
Rosalina
Discorbinella
Neoconorbina
Glabratella
Quinqueloculina spp.
Triloculina spp.
Sigmoilina grata (Terquem)
Sigmoilina costata Schlumberger
Massilina
Cibicides spp.
Planorbulina
Acervulina
Peneroplids
Amphistegina
Eggerella scabra (Williamson)
Trochammina squamata Parker and Jones
Haplophragmoides canariensis (d'Orbigny)

Zone II 50–150 m Preferred habitat of:

Textularia
Eponides repandus (Fichtel and Moll)
Gaudryina rudis Wright

Accompanied on muddy sediment by:
Nonionella turgida (Williamson)
Reophax scottii Chaster
Valvulineria bradyana (Fornasini)

In this zone *Eggerella scabra* (Williamson) and *Haplophragmoides canariensis* (d'Orbigny) disappear. New forms that appear are *Nonion parkeri* le Calvez, *Hyalinea balthica* (Schröter), *Lagenodosaria scalaris* Batsch, *Cassidulina laevigata carinata* Silvestri and *C. crassa* d'Orbigny.

Bulimina replaces *Bolivina*, and between 100 and 150 m *Bigenerina* replaces *Textularia*.

Zone III   150–200 m   Transition zone to deeper water; introduction of deeper water species:

Bigenerina nodosaria d'Orbigny
Bigenerina digittata d'Orbigny
Pseudoclavulina crustata Cushman
Siphotextularia concava (Karrer)
Adercotryma glomerata (Brady)
Hyperammina
Rhizammina
Bathysiphon
Pyrgo

Zone IV   200–1000 m   Development of the deeper water species first seen in the preceding zone. Benthic individuals form less than 50 per cent of the total assemblage (more than 50 per cent planktonic). New arrivals include:

Triloculina fischeri Schlumberger
Cyclammina cancellata Brady
Lingulina seminuda Hantken
Uvigerina mediterranea Hofker
Epistomina elegans (d'Orbigny)

Some groups increase in abundance with increasing depth, e.g. *Pyrgo* and some agglutinated species.

Uvigerina mediterranea Hofker reach its maximum between 400 and 1000 m (30–35 per cent at 800 m). The optima of *Hyalinea balthica* (Schröter) and *Lagenodosaria scalaris* Batsch are 200–600 m and 400–700 m respectively.

Zone V   Deeper than 1000 m   The benthic fauna remains unchanged but the total population is 95–99 per cent planktonic.

Comparison of the oceanographic data with the depth zonation suggests the following correlations:

Zone 1 is controlled by the depth of penetration of light and hence the development of the plants upon which the foraminiferids find habitats.

Zone 2 represents the surface water layer deeper than the limit of plant colonization.

Zone 3 corresponds with the base of the surface layer or the top of the intermediate layer. It lies more or less at the lower limit of winter vertical convection.

Zone 4 spans the intermediate water with its salinity maximum at 300–400 m. The lower part of the zone corresponds with the upper part of the transition zone.

Zone 5 spans the transition and deep-water layers where temperatures and salinities are very stable. The benthic foraminiferids of this zone do not differ from those of zone 4 but the planktonic/benthic ratio is much higher (more than 95 per cent planktonics) in the dead assemblages.

FAUNAL BOUNDARIES

The studies of Blanc-Vernet (1969) suggest that the benthic foraminiferid fauna is fairly homogeneous in the same environments and at the same depth zones throughout the northern Mediterranean with the exception of zone 1. Here submarine vegetation of phanerogams and algae provide habitats for a variety of species in the depth zone 0–40 m. In this shallow water seasonal temperature changes reach their maximum (greater than 9 °C and up to 14 °C) and the summer temperatures are subtropical.

In the eastern Mediterranean, *Amphistegina madagascariensis* d'Orbigny, *Sorites variabilis* Lacroix, *Peneroplis pertusus* (Forskål), *Peneroplis planatus* (Fichtel and Moll) and *Vertebralina striata* d'Orbigny are present. These forms are commonly found in similar environments in the Indian Ocean and are therefore considered to be subtropical.

Sorites and *Amphistegina* do not penetrate into the northern Aegean Sea because of unfavourable salinities. *Amphistegina* does not extend to the north of Peloponnisos (Figure 79) perhaps because of the summer temperatures (<26 °C). *Sorites* extends as far as the Côte d'Azur but not as far as Marseille. It is abundant around Peloponnisos and in the southern Aegean Sea. The limit of distribution is probably related to the 23 °C summer isotherm. Likewise the peneroplids and *Vertebralina* extend from the east to the Marseille region but they are absent from the Golfe du Lion and from the Spanish coast at least as far south as the Ebro Delta, although they occur in the Balearic Islands. The factors controlling their disappearance are probably unfavourable summer temperatures combined with the introduction of fresh water from the Rhône and the Ebro causing somewhat reduced salinities in the shallow nearshore zone.

Apart from these subtropical species, many of the smaller benthic species from the various depth zones are found in the North Atlantic along the western European seaboard. For the fauna in general, as long ago as 1844 Edward Forbes had noted that in the depth zone 0–60 m there were more tropical species, and in the depth zone 60–500 m more northern species, in the Mediterranean (Ekman, 1953).

9. Shelf Seas: European Seaboard

Studies of the Western Approaches to the English Channel and the Celtic Sea have been carried out by Murray (1965c, 1970c). The distribution of water masses in the area is believed to be complex, although there are few detailed studies. Cooper (1966, 1967) points out that, for the latitude, the

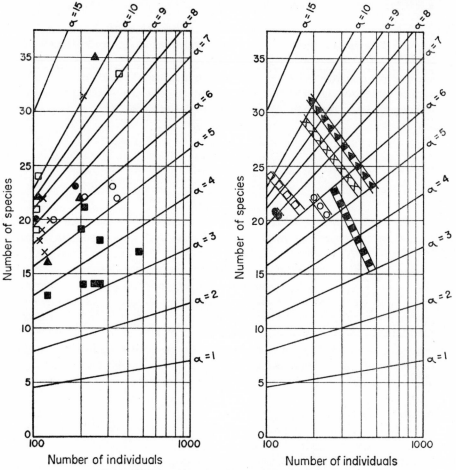

Figure 80 Diversity plot for the Western Approaches to the English Channel (from Murray, 1970c).
● = shelf edge, ○ = Celtic Sea, ▪ = Bristol Channel, ▲ = English Channel, X = South coast of Cornwall, □ = Eddystone–Plymouth.

climate is temperate. Gales and tidal currents control most of the water movements. There is normally little exchange of water between the Atlantic Ocean and the shelf, except as a result of southerly gales or as compensation currents replacing losses. The water on the continental slope at a depth of 800–1200 m is a mixture of waters derived from the Mediterranean and North Atlantic Central water. The surface water has a salinity of around 35 per mille, and the temperature varies from a winter mean of 9–10 °C to a summer mean of 16 °C. A thermocline develops in the summer, but mixing caused by storms commonly occurs. The sediments vary greatly according to the tidal currents.

The areas which have been studied include Plymouth and the south coast of Cornwall (Murray, 1965c, 1970c), South of the Lizard in the English Channel, Bristol Channel, Celtic Sea and shelf edge. The results are shown in Figures 80 and 81 and Table 18. Each of these areas is a distinctive en-

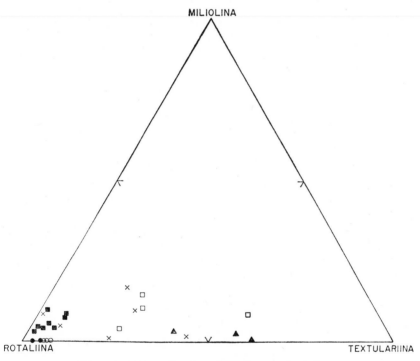

Figure 81 Triangular plot of Western Approaches to the English Channel. Symbols as in Figure 80.

vironment, and this is reflected in the differences between the foraminiferid assemblages.

The diversity of the Bristol Channel region is low ($\alpha = 3$ to $\alpha = 5$), and this may be due to the influence of fresh-water input from the Severn estuary and the rivers of South Wales. The English Channel samples are shell sands which represent a winnowed lag deposit. The dominant species (Table 18) are small in size, and most of them probably live clinging to shells on the substrate. This is a feature of areas subject to powerful currents.

Table 18 Summary of the Living Assemblages

	Shelf edge	Celtic Sea	Bristol Channel	English Channel	Cornwall–S. Coast	Eddystone–Plymouth
Dominant species	Cassidulina obtusa Cassidulina carinata Brizalina pseudo-punctata Brizalina spathulata Locally: Elphidium sp. Nonionella turgida Sta. 1557. Sparse living forms	Nonionella turgida Hyalinea balthica Bulimina marginata Locally: Cassidulina carinata Cassidulina obtusa Martinottiella communis	Bulimina gibba/elongata Fursenkoina fusiformis Locally: Quinqueloculina seminulum Cancris auricula Eggerella scabra	Gavelinopsis praegeri Cribrostomoides jeffreysii Spirillina vivipara Trochammina globerini-formis pygmaea Trochammina ochracea Locally: Cibicides lobatulus Patellina corrugata Textularia sagittula group	Ammonia beccarii Bulimina gibba/elongata Eggerella scabra Fursenkoina fusiformis Locally: Asterigerinata mamilla Gavelinopsis praegeri Nonion depressulus Quinqueloculina oblonga Textularia sagittula group	Ammonia beccarii Bulimina gibba/elongata Eggerella scabra Fursenkoina fusiformis Locally: Brizalina pseudopunctata Gavelinopsis praegeri Nonion depressulus Quinqueloculina oblonga Reophax fusiformis
Standing crop per 30 cm²	190 (1 sample)	205–342	122–472	0–249	—	—
Biomass: mm³ per 30 cm²	0.339 (1 sample)	0.966–2.413	0.395–103	0–0.657	—	—
Fisher α index	7	5–6	3–5	5–12	5–10	6–10
Similarity indices	45	69–85	38–89	61–64	19–64	37–67
No. of sp. common to sample pairs	12	13–16	6–11	11–21	7–14	9–12
Substrate	Muddy, fine sand	Muddy, fine to coarse, quartz sand with shell debris	Fine to coarse sands	Shell sands	Shell Variable	Fine sands
Depth (m)	420–1002	128–138	66–91	84–95	14–42	51–59

In the Plymouth region, samples have now been studied from April 1962, July 1962 and April–May 1968. A comparison of the dominant species is shown in Table 19. Changes are apparent from one season to

Table 19 Comparison of dominant species in seasonal samples
from the Eddystone–Plymouth traverse

	April 1962*	July 1962*	May 1968
Ammonia beccarii	X	X	X
Bulimina spp. = *B. gibba/elongata*	X		X
Triloculina spp. = in part *Quinqueloculina* oblonga	X	X	
Textularia sagittula group	X		
Reophax fusiformis	X		
Eggerella scabra	X		X
Fursenkoina fusiformis		X	X

* Data from Murray, 1965c

another, and this emphasizes that perhaps not too much significance should be attached to dominant species when only one set of samples is available from shallow shelf areas.

The differences between the living and dead assemblages in each of these areas is much greater than can be attributed to such seasonal causes. Similarity values between living and dead assemblages from the same stations show the following ranges:

	Similarity (%)	Species in common
Cornish coast	25–49	14–17
English Channel	7–9	2–8
Bristol Channel	30–67	7–14
Celtic Sea	48–60	13–16
Shelf edge	44–63	13–15

The dominant species and other data are summarized in Table 20. The least satisfactory similarity is between the living and dead assemblages of the English Channel shell sands. This must be attributed to the powerful currents. The small, living species cease to cling to the substrate on death and are either transported away or destroyed.

On the French side of the Channel, Moulinier (1967) found few living foraminiferids in the Baie de Seine.

Off the coast of Connemara, Ireland, in waters of nearly normal salinity, the substrates include seaweeds, sea-grass (*Zostera*), pebbles, shells and sand. Samples collected by Lees, Buller and Scott (1969) showed an absence of living foraminiferids on sedimentary substrates over much of the area. The maximum standing crop was 99 per 1.4 cm^3 sediment, but most samples yielded less than 20 per 1.4 cm^3 (=140 per 10 cm^3). *Cibicides lobatulus*

Table 20 Summary of the Dead Assemblages

	Shelf edge	Celtic Sea	Bristol Channel	English Channel	Cornwall–S. Coast	Eddystone–Plymouth

Dominant species	Sta. 1437 and 1428
	Brizalina spathulata
	Cassidulina carinata
	Cassidulina obtusa
	Trifarina angulosa
	  plus *Bulimina* cf. *B. alazanensis* at
	Sta. 1437
	Sta. 1557
	Cibicides lobatulus
	Textularia sagittula group
	Trifarina angulosa
		Bulimina marginata
		Cassidulina carinata
		Cassidulina obtusa
		Fursenkoina fusiformis
		Hyalinea balthica
		Nonionella turgida
		Textularia sagittula group
		Locally:
		  Trifarina angulosa
			Bulimina gibba/elongata
			Quinqueloculina seminulum
			Textularia sagittula group
			Cibicides lobatulus
			Locally:
			  Planorbulina mediterranensis
			  Eggerella scabra
			  Elphidium excavatum
			  Fursenkoina fusiformis
				Gaudryina rudis
				Textularia sagittula group
				Cibicides lobatulus
				Locally:
				  Quinqueloculina seminulum
					Ammonia beccarii
					Cibicides lobatulus
					Quinqueloculina seminulum
					Textularia sagittula group
					Locally:
					  Asterigerinata mamilla
					  Eggerella scabra
					  Elphidium crispum
					  Gavelinopsis praegeri
					  Planobulina mediterranensis
					  Quinqueloculina bicornis
					  Rosalina globularis
					  Quinqueloculina oblonga
						Ammonia beccarii
						Cibicides lobatulus
						Gavelinopsis praegeri
						Textularia sagittula group
						Locally:
						  Asterigerinata mamilla
						  Bulimina gibba/elongata
						  Eggerella scabra
						  Protelphidium anglicum
						  Quinqueloculina seminulum
						  Trifarina angulosa

Fisher α index	9–15	6–10	3.5–6.5	2.5–6.5	6–13	6.5–11
Similarity indices	4 69 —	52–81	25–89 mainly 50–90	32–89	31–71	46–80
No. of sp. common to sample pairs	22 —	16–22	9–16	7–14	13–29	18–31

(Walker and Jacob) is the dominant species. Seaweeds and sea-grass had *Discorbis columbiensis* Cushman and *Cribrostomoides jeffreysii* (Williamson) as additional dominant species. Four samples of more than 100 individuals give an α range of 1.5 to 3.5, 1–17 per cent Miliolina and 83–99 per cent Rotaliina. On pebbles the dominant species are *C. lobatulus*, *D. columbiensis* and *D. bradyi* Cushman; α = 2 to α = 4.5. Miliolacea are almost absent, Rotaliina 87–97 per cent, Textulariina 2–12 per cent.

Off the Île d'Yeu in the Bay of Biscay, Rouvillois (1970) studied samples from a depth range of 60–145 m. The sediments range from muddy to gravelly sands. The area is influenced by the North Atlantic Drift. Salinity is 35 per mille and temperature at the bottom is believed to vary from about 10 to 12 °C.

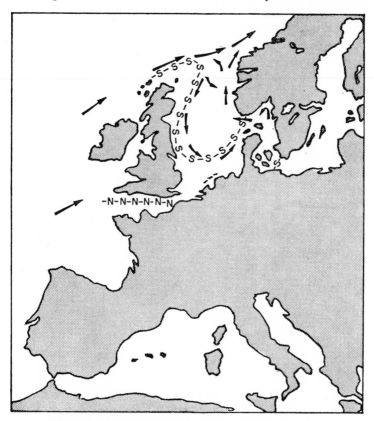

Figure 82 Faunal boundaries along the European seaboard (based on Murray, 1971). S-S-S = northern boundary of southern species, N-N-N = southern boundary of northern species.

The samples were collected during March and April 1968. Living individuals are rare, and only 25 of the 60 species are represented in the living assemblages. The standing crop is 2 to 10 per 10 cm². The principal forms are *Bigenerina cylindrica* Cushman, *Cancris auriculus* (Fichtel and Moll), *Nonion asterizans* (Fichtel and Moll), *Gavelinopsis praegeri* (Heron-Allen and Earland) and *Hyalinea bathica* (Schröter).

It seems probable that the low abundance of living individuals in this study is due to the method of sampling (dredge), as similar muddy substrates in the Celtic Sea have yielded much bigger standing crops.

Haman (1972) recorded patchy distribution of living forms in Tremadoc Bay, Irish Sea. Most of his data concerns total (living plus dead) assemblages.

Murray (1971) recognized three groups of inner shelf species along the European seaboard: those which are widely distributed, those of southern origin which reach the northern limit of their distribution around Britain, and those of northern origin which reach their southern limit of distribution (Figure 82). Thus Britain spans a diffuse faunal boundary between northern and southern influences, and this is controlled by the water circulation. In particular, a zone of homohaline and homothermal water extends along the east coast of Britain and across the southern North Sea to Denmark. It enables favourable temperature conditions to exist on the sea floor during the summer, which in turn enables the southern foraminiferids to reproduce. The remainder of the North Sea is subject to thermal stratification, so conditions on the sea floor are never warm enough for southern species. In this region Thorson (1957) has recognized a boreal community comprising *Astrorhiza arenaria* Carpenter, *Saccammina sphaerica* Sars, and *Psammosphaera fusca* Brady together with *Thyasira* and a few *Amphineura*.

10. Shelf Seas: Sundry Areas— Japan and Antarctica

JAPAN

Twenty-six Phleger cores from Miyako and Yamada Bays on the north-east coast of Japan were found by Ujiié and Kusukawa (1969) to contain sparse living assemblages. Only 4 samples yielded more than 100 individuals. *Eggerella advena* (Cushman), *Hopkinsina pacifica* Cushman and *Cassidulina complanata* Ujiié and Kusukawa are the dominant species. The authors commented on the big difference in composition between the living and dead assemblages.

Six stations at depth 15–97.5 m in Tanabe Bay were sampled by Uchio (1967) using a dredge. The only observations on living foraminiferids concerns the occurrence of *Operculina ammonoides* (Gronovius) and *Hanzawaia nipponica* Asano at a depth of 50 m on fine sand. Further sampling by Chiji and Lopez (1968) showed that living individuals were present to the extent of below 50 to more than 200 per 20 g sample. Most of their abundances are estimated and there are only fourteen samples. However, they consider that the following distributions are noteworthy:

Ammomarginulina foliaceus (Brady) and *Ammonia beccarii* (Linné) var. *tepida* (Cushman) are widely distributed in the bay. *Amphistegina radiata* (Fichtel and Moll) and *A. madagascariensis* d'Orbigny occur in the open sea and at the bay mouth, in association with *Operculina ammonoides* (Gronovius) and *Textularia articulata* d'Orbigny. In the inner bay, *Ammonia beccarii* (Linné), *Haplophragmoides canariensis* (d'Orbigny), and *Nonionella miocenica* Cushman are characteristic. Diversity throughout the bay ranges from α = <1 to α = 2.

Tanabe Bay has salinities of 34 to 35 per mille, and bottom water temperatures of 19.2 to 21.5 °C were recorded at the time of sampling. The very low diversities here are anomalous and may be due to the method of estimating the species abundances.

Ishikari Bay on the west coast of Hokkaido is a region of very variable sediment type (mud to gravel grades). The oceanographic conditions are complex and are controlled by the interplay of three water masses: coastal water, upwelled cold water and the Tsushima warm water current. Ikeya (1970) collected samples from depths of 10–500 m. Eleven samples yielded standing crops greater than 100 individuals (total range 0–608 per 10 cm²). The diversity ranges from α = <2 to α = 4, and there is no variation with depth. On a triangular plot (Figure 83) the samples show little correlation between depth and composition, although most are rich in Textulariina and contain no Miliolina. The exception is station 30 with 37 per cent Miliolina.

This comes from a depth of 60 m, outside the bay proper. The substrate is sand, and near normal salinities prevail. The fauna is dominated by *Cibicides refulgens* Montfort, *Hanzawaia nipponica* Asano, *Pseudononion japonicum* Asano, and *Quinqueloculina seminulum* (Linné).

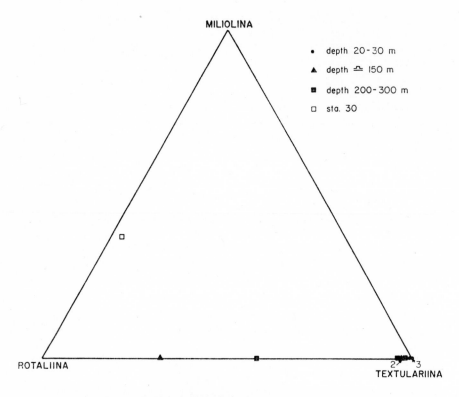

Figure 83 Triangular plot of samples from Ishikari Bay, Japan (data from Ikeya, 1970).

The four samples, from 20–30 m depth, are dominated by *Eggerella advena* (Cushman), *Lagenammina difflugiformis* (Brady), *Trochammina charlottensis* Cushman and *T. pacifica* Cushman. At the shelf edge, at a depth of roughly 150 m, the principal species are *Goësella flintii* Cushman and *Trochammina charlottensis* Cushman, although several other species are common at individual stations. In the 200–300 m depth range, *Alveolophragmium* cf. *A. subglobosum* (Sars), *Eggerella advena* (Cushman), *Haplophragmoides bradyi* (Robertson), *Textularia earlandi* Parker and *Trochammina quadriloba* Höglund are the dominant forms.

Ikeya used a geometrical series of population analysis to demonstrate that the death assemblages are markedly different from the living. He suggested that the cause of the difference is mixing of the assemblages after death.

ANTARCTICA

The following foraminiferids believed to have lived at a depth of 10–40 m, clinging to holothurians and algae, were recorded by Blanc-Vernet (1965) from near Adelie Land:

Pelosina variabilis Brady
Marsipella cylindrica Brady
Hyperammina sp.
Hippocrepina indivisa Parker
Jaculella acuta Brady
Textularia earlandi Parker
Trochammina ochracea (Williamson)
Trochammina malovensis Heron-Allen and Earland
Trochammina antarctica Parr
Cibicides lobatulus (Walker and Jacob)
Cibicides refulgens Montfort

C. lobatulus was conspicuously abundant on some holothurians.

11. *Shelf Seas: Shallow Land-locked Seas*

Seas which are to a large extent surrounded by land are more variable environments than open seas. Three broad groups can be recognized: in regions where runoff from the land exceeds evaporation from the sea, the waters are hyposaline, e.g. Hudson Bay, the Baltic Sea and the Black Sea; in regions where runoff from the land is small compared with evaporation from the sea, the waters are hypersaline, e.g. the Persian Gulf; and the normal marine, Limski Channel. In a sense land-locked seas could be regarded as very large lagoons.

HYPOSALINE

Hudson Bay

This is the largest enclosed sea in North America. The area is about 520 000 km². The average depth is 90 m, and the deepest part is 280 m (Leslie, 1965). Depths greater than 200 m are confined to a basinal area in the central and northern part of the Bay. This leads to the Hudson Straits and is believed to represent a submerged valley system. Variation in bottom topography around the margins is related to the bed rock geology.

The bottom water has a salinity of 32–33 per mille and temperatures of 1 to 1.5 °C. The surface 30 m are affected by seasonal temperature changes (temperatures up to 9 °C) and by runoff from rivers leading to hyposaline conditions. In the winter, ice covers the entire Bay.

The circulation is anticlockwise, with south-flowing currents on the west side and north-flowing currents on the east. Tides are greatest in the west (4–6 m) and least in the east (0.5–1 m).

Leslie (1965) studied grab samples. His method of analysis were different from those of most authors so that it is not possible to replot his data. From each grab sample, he took part of the surface 2 cm and, after staining, sieving and drying, he calculated the foraminiferal number and the number of living individuals per gram of dry sediment. The foraminiferal number is analogous to the total population, and the number of living individuals per gramme of sediment is analogous to the standing crop. For the latter, values of 0.1 to 16.2 per gram from Hudson Bay were said to be similar to those on the continental shelf off southern California obtained by Bandy, Ingle and Resig (1964a). The highest values were obtained along the northern, western and south-western coastal areas of the Bay, where it is believed that nutrients are contributed by the rivers, that sedimentation is slow, and that the bottom water has high dissolved oxygen. By contrast, the areas of low living values (the central bay and east coast) have deeper water, faster sedimentation, poor food supply and lower oxygen.

Of the 92 species encountered, 13 were dominant in the total populations, although only 11 were dominant in the living condition. Leslie analysed the environmental parameters measured at the stations of abundance for each of these species. Most showed some preference for certain depth zones and substrate types, and to a lesser extent for temperatures, salinity and oxygen content of the water. A depth zonation was apparent:

1. *Shallow bay* 26–130 m *Eggerella advena* (Cushman) (well sorted fine sand) and *Protelphidium orbiculare* (Brady) (good oxygen, 40 per cent sand substrate).

2. *Intermediate bay* 50–175 m *Textularia contorta* Höglund (low oxygen, sand, silt or clay), *Spiroplectammina biformis* (Parker and Jones) (no substrate preference), and *Buccella frigida* (Cushman) (high oxygen, silty sand or sandy silt).

3. *Deep bay.* 100–230 m—*Cassidulina norcrossi* (Cushman) (silty clay, clayey silt), *Recurvoides turbinatus* (Brady) (clay or clayey silt), *Adercotryma glomeratum* (Brady) (no substrate preference) and *Melonis zaandamae* (Van Voorthuysen) (silty clay or clayey silt). Additional common dead species are *Cassidella complanata* (Egger) and *Buliminella elegantissima* (d'Orbigny).

In addition, *Elphidium incertum* (Williamson) (clayey silt) and *Cassidulina islandica* Nørvang (the most abundant species with no clear environmental controls) are present at all depths. A conspicuous feature of the assemblages is the absence of Miliolina, probably due to the low salinities and low temperatures.

Baltic Sea

Due to fresh-water runoff from rivers, there is a surface layer of 40–80 m thickness of fresh water which flows seawards, while sea water of 34 per mille salinity enters the Baltic along the bottom. Thus under normal conditions the waters are stratified. Along the plane of junction of the two water bodies some mixing takes place, and the surface waters increase in salinity from east to west near the entrance to the Baltic (Lutze, 1965). In the Beltsee the surface waters reach a salinity of 10–15 per mille but during periods of storms vertical mixing takes place and the water column becomes isohaline. Segerstråle (1957) has pointed out that the salinity pattern of the Baltic is very stable compared with that of estuaries.

The Baltic is divided into a number of deeper basins separated by sills. The incoming denser sea water flows along the bottom into the first basin. Some flows over the sill into the next basin and so on, but with passage further into the Baltic the salinity of the bottom water decreases (Arkona Basin 15 per mille, rarely up to 20 per mille, Bornholm Basin 13–16 per mille, Gotland Basin 12.5 per mille, Landsort Deep 11 per mille) (Lutze, 1965).

Thus three main water masses can be recognized: inflowing North Sea water (34 per mille salinity), surface waters (hyposaline), and basin waters (hyposaline). Of these, the surface waters show the greatest variation, not only with respect to salinity but also to temperature. Further, during periods of poor circulation, oxygen consumption in the bottom waters causes the deeper parts of the basins to become anoxygenic.

Most of the sediments shallower than 20 m are sandy. At greater depths silt and mud predominates, except in areas subject to powerful currents.

Lutze (1965) has made a detailed study of the living foraminiferids. The following list compares the species names used by Lutze with those used here:

This book	*Lutze (1965)*
Elphidium excavatum (Terquem)	*Cribrononion excavatum* (Terquem)
Elphidium articulatum (d'Orbigny)	*Cribrononion* cf. *alvarezianum* (d'Orbigny)
Elphidium cf. *E. gerthi* Van Voorthuysen	*Cribrononion* cf. *gerthi* Van Voorthuysen
Elphidium incertum (Williamson)	*Cribrononion incertum* (Williamson)
Elphidium asklundi Brotzen	*Cribrononion asklundi* (Brotzen)

Most of Lutze's results are expressed as cumulative histograms drawn along sampling traverses. He did not give tables of data so that it is not possible to determine the α values. However, he counted 200–500 individuals from each sample and he recorded a total of 25 species. Even if all the species occurred in one assemblage of 200 individuals (which is unlikely), the diversity would only be $\alpha = 7.5$. This is the highest value possible from these figures, so clearly most samples had $\alpha < 7$. Lutze comments (1965, p. 132) that the number of species decreases with increasing brackishness.

Lutze presented his results in a series of traverse profiles showing the relationship of the living and dead assemblages, standing crop, sediment composition and depth. He concluded that the distribution of the foraminiferids is controlled by the three main water masses: surface water-layer brackish fauna, marine fauna and arctic-boreal brackish fauna.

The surface water-layer brackish fauna has great affinity with the shallow water assemblages of the North Sea. Typical species are *Miliammina fusca* (Brady), *Elphidium asklundi* Brotzen and *Elphidium* cf. *E. gerthi* Van Voorthuysen. Locally common forms include *Elphidium articulatum* (d'Orbigny), *E. excavatum* (Terquem), *Asterellina pulchella* (Parker) and *Ammoscalaria runiana* (Heron-Allen and Earland). The surface water shows a wide range of salinity (5–19 per mille) and temperature (0–19 °C; see Figure 84). Variation in sediment type is believed not to influence the occurrence of individual species. However, a flat form of *E. excavatum* is found in the influence of surface waters, while an inflated form is characteristic of the brackish water of the deeper parts of the basins. The standing crop is commonly less than 10 per 10 cm², but there are occasional values greater than 200 per 10 cm².

The marine fauna is found in the region of inflowing more saline water from the North Sea. Salinity is 17–29 per mille and temperature is 1–12 °C (Figure 84). Typical species include *Elphidium incertum* (Williamson), *Ammotium cassis* (Parker) and *Ammonia beccarii* (Linné), *Eggerella scabra*

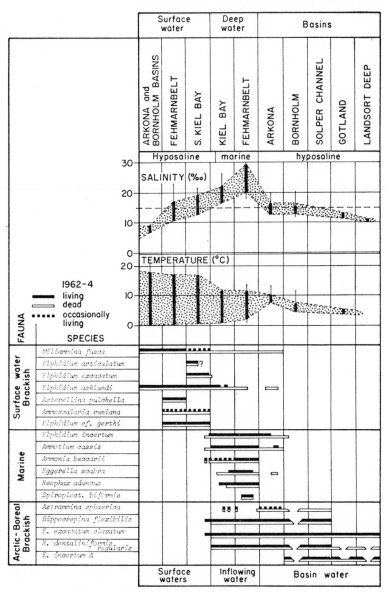

Figure 84 Distribution of Baltic Sea foraminiferids
(redrawn from Lutze, 1965).

(Williamson), *Reophax aduncus* Brady and *Spiroplectammina biformis* (Parker and Jones). The standing crop varies from 0 to 250 per 10 cm². Lutze considers that only about 30 per cent of the species are found also in the southern part of the North Sea. The remainder come from the Arctic and the deep, cold, waters of the Atlantic.

The arctic-boreal brackish water fauna is found in the water of the deeper parts of the basins. Salinity and temperature are not very variable (10–15 per mille and 5–8 °C respectively). An inflated form of *Elphidium excavatum* (Terquem) (var. *clavatum* of Lutze) occurs, together with species which are not found in the North Sea but only in the Arctic or deep Atlantic: *Astrammina sphaerica* (Heron-Allen and Earland), *Hippocrepina flexibilis* (Wiesner), *Reophax dentaliniformis regularis* Höglund and *Elphidium incertum* (Williamson) subspecies A. The standing crop is normally very low (less than 5 per 10 cm²) because of the oxygen deficiency. For the same reason, most of these species live not in the deepest parts of the basins but around their edges, close to the transition with the overlying oxygenated waters.

In each of these assemblages miliolids are absent and there are mixtures of Rotaliina and Textulariina. Thus the data from the Baltic accord with those from other hyposaline environments.

As might be expected in a region subject to variations in water circulation, oxygen supply, temperature and salinity, there is a variation in the distribution of individual species from one year to another. This is reflected in the difference between the living and dead assemblages. *Ammoscalaria runiana* (Heron-Allen and Earland) and *Miliammina fusca* (Brady) are abundant in the dead assemblage of the Gabelsflach area of Kiel Bay. However, they were not living there at the time of sampling. Similar variations have been found for *Ammonia beccarii* (Linné). Solution of calcareous shells is thought to take place in the Arkona Basin, and in the Bornholm Basin many of the shells (including those still containing some protoplasm) are etched. Lutze considered it possible that *Elphidium excavatum clavatum* could outlast short unfavourable periods by contracting the protoplasm from the outer three or more chambers (1965, p. 128). Other changes in the dead assemblages are brought about by the transport of tests from shallow to deeper water, i.e. from the swells and shallows into the basins.

Lutze suggested that during the last thirty years, *Ammotium cassis* (Parker) and *Reophax aduncus* Brady have increased in abundance. This is attributed partly to adaptation and partly to a slight increase in salinity. This is one of the few long-term changes recorded in the literature.

North-west Black Sea

Macarovici and Cehan-Ionesi (1962) used an unusual method to assess the abundance of living foraminiferids: they dissolved the tests in dilute acid and stained the residue with paracarmine. In this way they determined the percentage of living individuals in the total assemblage. However, they did not establish whether all the species were represented by living individuals, although they seem to have assumed this to be so.

In the depth range 24–100 m, in the north-west Black Sea on muddy sand, muddy shell sands, and hyposaline water (salinity 15–18 per mille) the living assemblage comprises:

> *Ammonia beccarii* (Linné)
> (= *Rotalia beccarii* of Macarovici)
> *Discorbis villardeboana* d'Orbigny

Nonion depressulum (Walker and Jacob)
Nonion stelligerum d'Orbigny
Elphidium poeyanum d'Orbigny
Eggerella scabra (Williamson)
(=*Verneuilina scabra* of Macarovici)
Quinqueloculina seminulum (Linné)

Although their data cannot be plotted on α or triangular diagrams, it is clear that the α values must be low because of the small number of species encountered, and the Miliolina are an insignificant component.

HYPERSALINE: PERSIAN GULF

A general account of the area has been given by Evans (1966b). The hot climate and small runoff cause the sea to become hypersaline. Salinity increases from 37–38 per mille at the entrance to 42 per mille along some of the coastal areas. In lagoons, salinities rise much higher. Throughout the greater part of the Gulf, the modern sediments are carbonate muds with varying amounts of biogenic debris.

Murray (1966c) examined samples collected on three traverses on the shelf by the Trucial Coast. Although the samples were preserved in alcohol, few living foraminiferids were encountered. It is possible that most of the living individuals would be found on weeds (which were not sampled), as is the case in the lagoons (see Murray, 1970b). This idea receives support from the results of Haake (1970) who found, on average, one living *Quinqueloculina* in every 500 dead tests in samples from the axial region of the Gulf.

A general decrease in standing crop of the macrofauna of both muddy and sandy substrates with passage from the entrance to the head of the Persian Gulf was recorded by Thorson (1957). The diversity of the microfauna seems to be less than that of the adjacent Indian Ocean shelf seas.

This type of environment needs to be investigated much more because of its great geological importance.

NORMAL MARINE: LIMSKI CHANNEL

The Limski Channel at Rovinj, Jugoslavia, has been studied in detail by Daniels (1970). The channel is less than 1 km wide, 11 km long and for much of its length it has a depth of 34 m. It receives little fresh-water runoff and, as it has an open connection with the Adriatic, salinities remain nearly normal at 35–38 per mille throughout the year. Temperature shows a cycle from coolest in winter (9 °C) to warmest in summer (25 °C). A thermocline develops in May and persists until the autumn. During the winter, vertical mixing takes place. The substrate is muddy silt burrowed to a depth of 30 cm.

Foraminiferid samples were collected monthly from eleven stations over a fourteen-month period. There are two dominant species: *Ammonia beccarii* (Linné) in the landward part and *Nonionella opima* Cushman in the seaward part. Both form more than 20 per cent of the living assemblage. Daniels recognized five zones from the mouth to the head of the channel based on small differences of species distributions.

Standing crop values are very high. The average for each station is in the range 400–800 per 10 cm². The variation at individual stations from month to month are great, and at one station ranges from 200 to 1000. In general there is a decrease in standing crop from the mouth to the head.

Diversity data calculated from Daniel's data show generally high values of α. The seasonal range at each station is shown in the following table:

Mouth										Head
Sta-										
tion	39	11	5	27	9	8	17	36	52	1	51
α	11–18	13–19	10–15	10–15	9–15	10–14	9–14	7–13	6–13	7–13	9–14

Details of the variation throughout the year at each station are summarized graphically in Figure 85. From the mouth to station 17, where depths are greater than 20 m, there is an increase in diversity in June–August and

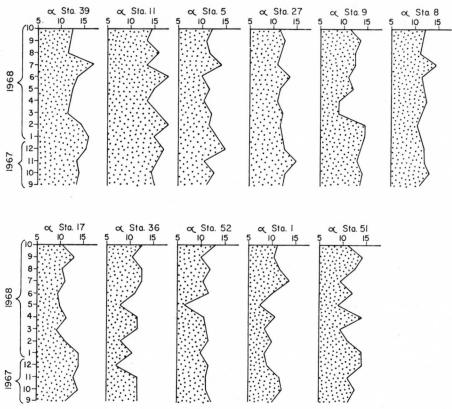

Figure 85 Annual variation in α diversity values at stations in the Limski Channel (data from Daniels, 1970).

December–January. Lower values are found in March–May. From station 36 to the head, in depths greater than 20 m, there is a less consistent pattern except that month 5 (May) is always low. Taking the monthly values, on average there is a slight decrease in α from the mouth to the head.

Daniels discussed in detail the seasonal changes in the triangular plots. He points out that the greatest influence is due to the seasonal changes in the abundance of the main species, particularly *Nonionella opima* Cushman. This species is most abundant in the seaward end of the channel in September and October, and least abundant in November–March. Miliolaceans are most abundant in April, June and July, particularly at the head of the channel. The total field occupied by all samples is shown in Figure 86,

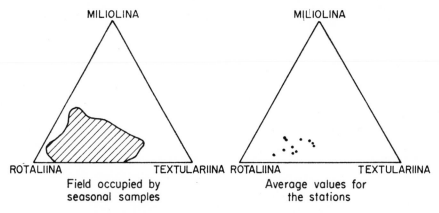

Figure 86 Triangular plots of Limski Channel data (based on Daniels, 1970).

together with the mean values for each station. Although the total variation is large, it is consistent with the overall field for shelf seas.

Samples taken from two cores show that foraminiferids can live in burrows down to a depth of 20 cm and the standing crop may reach up to eight to ten times that at the surface. This raises problems of the significance of standing crop data based on 1 cm-thick surface samples.

Limski Channel clearly offers a very favourable habitat for foraminiferids because of its stable environment which is not subject to powerful currents or climatic extremes. For these reasons diversity is very high.

12. Shelf Seas: Depth Zonation

The majority of papers concerning the distribution of shelf foraminiferids show a primary concern with depth zonation. This manifests itself in tables of data arranged according to the depth, rather than the position or number of the samples, and in cumulative histograms of traverses taken perpendicular to the shore. Since the studies are usually carried out by geologists, it is natural that they should seek modern distribution patterns that will be helpful in the interpretation of depth relationships of fossil deposits (see Funnell, 1967). However, there is the great disadvantage that objectivity may be lost in the process.

Phleger (1965c) has suggested that depth zones should be based on the shallowest occurrence of species, deepest occurrence of species and on their abundance. A simple and efficient way of measuring these is the similarity index. This takes account of the presence or absence of individual species in a sample pair. For species common to both samples the lowest relative abundance contributes to the index (page 12). Thus, faunal breaks should be revealed by low similarity (similarity index of less than 40 or 50 per cent). Almost identical samples show a similarity of more than 80 per cent, and moderately similar samples show an index of 50–80 per cent. This method has been used to re-interpret Uchio's depth zones for the Californian shelf (see pages 97–99).

In most studies the environmental data collected typically comprise depth and sediment type. Some authors provide information on temperature and salinity of the bottom water, but this is the exception rather than the rule. In any one study area, depth is usually the most variable environmental parameter that is measured. However, is depth itself an important parameter limiting the distribution of foraminiferids?

An idealized depth zonation is presented in Figure 87. The six depth zones are arranged parallel with the shore. Each could be characterized by a number of species confined to that zone, although other species would span several zones. Further, there would be a change in diversity from low in zone 1 to high in zones 3 and 4, and then possibly falling in zones 5 and 6. On the triangular plot we would expect zone 1 to comprise Rotaliina, Miliolina and Textulariina, zones 2–4 to be dominantly Rotaliina with some Textulariina and hardly any Miliolina, while zones 5 and 6 would see an increase in Textulariina. Finally we might expect the sediment to be sand in zone 1, silty sand in zone 2, silty mud in zone 3, and mud in zones 4–6.

Unfortunately such an arrangement is rare in modern shelf seas. In considering the reasons why this is so the following points are important:

1. Many shelf sea floors are carpeted with sediment that is not in equilibrium with present-day conditions. During the Pleistocene, lowered sea levels caused the accumulation of sands out as far as the shelf edge. Where

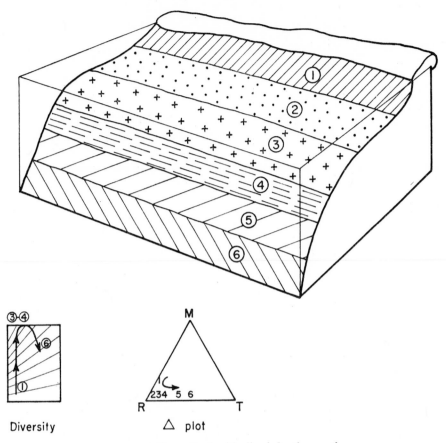

Diversity   △ plot

Figure 87 An idealized depth zonation on a continental shelf.

sedimentation is slow these relict sediments have not yet been buried by sediment in equilibrium with the modern sea-level.

2. Depth in itself conceals a variety of environmental parameters which individually may be important controls of foraminiferid distribution: hydrostatic pressure, light penetration, change in food supply from algae (including diatoms) to non-photosynthetic organisms deeper than the photic zone, salinity stratification, temperature stratification, dissolved oxygen, relative solubility of carbonates, and currents. All these are potential limiting factors, although in any one area only one or two will reach the critical limit for individual species. This means that in different areas of sea floor the distribution of individual species is controlled by different limiting factors, e.g. in one area temperature may be unfavourable for reproduction, while in another area temperature may be at the optimum but perhaps salinity is too low. Thus the presence of a species is dependent on all the factors which limit its life processes being within the limits tolerated. If any one critical factor exceeds the toleration, the species will not survive and will therefore be absent.

Table 21 Comparative depth ranges of living examples of selected species (depths in m)

	San Diego Uchio (1960)	Todos Santos Bay Walton (1955)	Gulf of California		Overall range	Reliability
			Phleger (1964a)	Phleger (1965c)		
Bolivina acutula Bandy	0–110	—	18–73	10–110	0–110	Good
Bolivina vaughani Natland	0–200	0–460	18–130	25–350	0–460	Poor
Bulimina denudata Cushman and Parker	36–460	36–460	18–130	—	18–460	Good
Buliminella elegantissima (d'Orbigny)	0–150	0–550	0–820	10–200	0–820	Poor
Cassidulina tortuosa Cushman and Hughes	36–150	18–90	—	—	18–150	Poor
Cibicides fletcheri Galloway and Wissler	0–120	0–460	—	—	0–460	Poor
Cibicides mckannai Galloway and Wissler	55–460	36–180	60–1500	75–200	36–1500	Poor
Epistominella smithi (R. E. and K. C. Stewart)	460–104	365–1100	—	—	365–1100	Good
Hanzawaia nitidula (Bandy)	36–130	0–90	18–360	15–200	0–360	Poor
Textularia schencki Cushman and Valentine	55–90	0–180	18–73	10–75	0–180	Poor

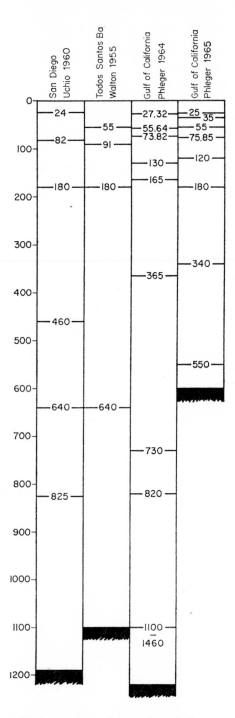

Figure 88 Comparison of depth boundaries proposed for the Pacific seaboard of North America.

In Table 21, species which show a restricted depth zonation in one of the four study areas in the California–Gulf of California region are listed with their recorded depth ranges. A summary column shows the overall range. Where the depth range in individual areas is close to the overall range, the reliability of the species as a depth indicator is good. Where there are big differences, it is poor. Only three of the ten examples are good. This clearly indicates that depth is not itself an important control except on the scale of tens of metres versus hundreds of metres. If *Buliminella elegantissima* (d'Orbigny) can live down to 550 m in Todos Santos Bay and 820 m in the Gulf of California, but only down to 150 m off San Diego, a factor or factors other than depth must be the control.

This strongly suggests that depth zones recognized in one area on the basis of certain indicator species are only applicable to that area and to adjacent areas where the environmental conditions are essentially the same.

A comparison of the depth boundaries recognized by four authors (Figure 88) suggests that the general depths at which change takes place (regardless of which species are used to define the change) are approximately 25 m, 80 m, 180 m and 350 m. At greater depths it is more difficult to find general agreement. In a general sense these subdivisions are nearshore shelf (25 m) inner shelf (25–80 m), outer shelf (80–180 m) and upper slope (180–350 m). Similar broad depth categories were used by Bandy (1964) in a useful review of depth distributions. At the present state of knowledge, only these broad divisions can be reliably recognized over large areas.

13. Deep-sea Foraminiferids

The limit of the ocean basins is conventionally placed at the edge of the continental shelf. At a depth of 180–200 m on most shelves, the inclination of the sea floor increases and this corresponds with the continental slope. Hedgpeth (1957a), in a review of marine environments, recognized a bathyal zone extending from the shelf edge to a depth of 4000 m, an abyssal zone from 4000 to 6500 m and a hadal zone extending from 6500 m to the deepest parts of the ocean (>10 000 m). The average depth of the ocean is 3800 m.

Sverdrup, Johnson and Fleming (1942) summarize the data on deep-sea salinity, temperature and oxygen concentration. The salinity is usually slightly less than 35 per mille, the temperature is in the range 1 to 4 °C at depths greater than 2000 m, and dissolved oxygen varies from less than 1 to more than 6 cm³/l. Compared with shallow waters, these conditions are extremely stable.

One important feature of the oceans is that, with increasing depth and increasing carbon dioxide concentration, the solubility of calcium carbonate increases. A critical level is the calcium carbonate compensation depth which lies at 4000 to 5500 m in different parts of the ocean. At greater depths, solution of calcareous shells takes place.

The most comprehensive record of deep-sea foraminiferids is that produced by Brady from the Challenger samples (Brady, 1884).

Few studies of living deep-sea foraminiferids have been carried out. Some of the data from the Californian seaboard relate to the bathyal zone but, because of their similarity with the shelf edge assemblages, they have not been separated off (see Chapter 5).

The Peru–Chile trench was investigated by Bandy and Rodolfo (1964). Samples were collected by trawl and by Phleger corers, from depths of 796 to 6011 m. Living foraminiferids occurred in seven of the thirteen surface sediment samples, and comprised the following species:

Bolivina costata d'Orbigny	796 m
Bolivina spissa Cushman	796 m
Bolivinita minuta (Natland)	796 m
Cassidella seminuda (Natland)	796 m
Cassidulina delicata Cushman	796 m
Cassidulina neocarinata Thalmann	1932 m
Cibicides wuellerstorfi (Schwager)	1932 m
Epistominella exigua (Brady)	1932–4606 m
Epistominella pacifica smithi (Stewart and Stewart)	796 m
Globobulimina pacifica Cushman	1932 m
Hoeglundina elegans (d'Orbigny)	1932 m
Nonion barleeanus (Williamson)	1932–4606 m
Pseudoeponides tener stellatus (Silvestri)	796–3257 m
Bulimina rostrata Brady	1171 m

Bulimina subacuminata Cushman and Stewart	1932 m
Cassidulina cushmani Stewart and Stewart	3257 m
Cassidulina subglobosa Brady	1932 m
Chilostomella ovoidea Reuss	1932 m
Bulimina affinis d'Orbigny	1932 m
Bulimina barbata Cushman	1932 m
Cassidella complanata (Egger)	1932–4606 m
Chilostomella czizeki Reuss	1932 m
Epistominella levicula Resig	1932–5929 m
Eponides tumidulus (Brady)	1932–6011 m
Gyroidina marcida (Emiliani)	1932 m
Pullenia bulloides (d'Orbigny)	1932–6011 m
Rotamorphina laevigata (Phleger and Parker)	1932 m
Nonion pompilioides (Fichtel and Moll)	3257–4606 m
Stilostomella antillea (Cushman)	6011 m
Uvigerina proboscidea Schwager	4606 m

All these species are representatives of the suborder Rotaliina. Nevertheless, the concentration of empty calcareous tests per gram of sediment falls from thousands at depths of less than 1000 m to 200 at depths greater than 4000 m. Bandy and Rodolfo proposed a series of bathymetric zones based on the total foraminiferid assemblages, but this is scarcely possible on the data for the living assemblages.

In a study of deep-sea meiobenthos from the Atlantic off North Carolina, Tietjen (1971) recorded the standing crop of foraminiferids. The values ranged from 1 to 62 per 10 cm^3 in the depth range 50 to 2500 m. The shelf was poor, 1–14 per 10 cm^3, compared with the slope, 1–62 per 10 cm^3. On the shelf, foraminiferids formed 2–22 per cent of the total meiofauna compared with 2–84 per cent in the deep sea. The top 1–2 cm of sediment was occupied by 95 per cent of the meiobenthos.

An unusual feature of the area is that in the depth range 800–1500 m, the zone of maximum organic content of the sediment, there are considerable quantities of salt-marsh grass and insect remains together with *Thalassia*. These are transported from the shore and deposited by the Western Boundary Undercurrent. They serve for food either directly or indirectly by providing a substrate for bacteria.

During 1966, Saidova (1967a) collected samples from the Kurile–Kamchatka Trench area, and obtained living foraminiferids from the depth range 250 to 9550 m. Twenty samples yielded 145 species. Saidova noted that the protoplasm of fresh, large species is greenish-grey. Living individuals show an irregular distribution with regard to depth, with greatest numbers at 1000–2000, 3000–4000 and 5500–8000 m. At these depths the standing crop is 0.55–1 per 10 cm^2 (=55 000–100 000 per m^2). At the intermediate depths the standing crop is less than 0.35 per 10 cm^2 (=35 000 per m^2) with particularly low values of less than 0.05 per 10 cm^2 (=5000 per m^2) at 4000–5000 m. Below 8000 m depth, the standing crop is less than 0.02 per 10 cm^2 (=20 000 per m^2). The biomass was calculated from the volume and an assumed specific gravity of 1.027 g/cm^3. Where the standing crop is highest, it reaches values of 4–10 g/m^2 and elsewhere 2–3 g/m^2. However, between 3000 and 6000 m there are large agglutinated species such as

Rhabdammina, *Astrorhiza*, *Hyperammina* and *Hormosina* so that, although the standing crop is small, the biomass reaches 8 g/m². These values are very small compared with those known from shallow-water environments.

Calcareous species dominate the assemblages to a depth of 2000 m, and below that agglutinated forms predominate. Saidova noted that all species were represented by both living and dead individuals.

Transport of shallow-water species into deep-water species has been recorded by Lukina (1967), and this emphasizes the need to use rose Bengal to differentiate living from dead individuals. Nevertheless, the subdivision of deep-sea faunas into taxocoenoses (based on genera and families) and genocoenoses (based on genera) from the study of dead individuals is reproduced here (Figure 89 based on Saidova, 1967b). The three principal calcareous taxocoenoses of the Pacific Ocean are:

1. Asterigerinidae
2. Buliminida
3. Alabaminidae

In the bathyal zone there are taxocoenoses of *Trochammina* and *Cyclammina*. In the abyssal zone there is a single taxocoenosis of Astrorhizida–Ammodiscida throughout the Pacific Ocean.

A characteristic feature of deep-sea deposits is the presence of large foraminiferids such as *Rhabdammina*, *Cyclammina*, *Pyrgo*, *Valvulineria* etc. The Peru–Chile Trench examples range in size from 1 to 10 mm. Deep-water forms of *Cyclammina* are larger than shallow-water representatives. Banner (1970) has reviewed the dead and living occurrences of this species and has concluded that the main limiting factors are *in situ* density and temperature. Most records of *C. cancellata* are from localities having a sigma-*t* value of around 27.7. Banner has stressed the need for observations of environmental parameters at the time of collection, especially now that new information is coming forward of deep-sea currents and contour-currents (see Welander, 1969).

In the summary of the scientific results of the 'Challenger' expedition, Murray (1895) commented on the large number of species observed in single deep-sea hauls. He also noted that there was a general decrease in the number of species and in the number of animals per unit sample with distance from shore and increase in depth.

Recently Hessler and Sanders (1967) used a special sampler to obtain reliable samples of the deep-sea fauna. Their results confirmed Murray's observation on the high diversity of the fauna and the reduction of standing crop with increased depth. They attributed the high diversity to the long-term 'climatic' stability of the deep-sea fauna.

Bandy and Rodolfo (1964) suggest that deep-sea foraminiferids evolved from two different groups which are related to the present distribution of species—one group includes forms which extend from shallow to deep water; these may have evolved in shallow water. The second group comprises species confined to the deep sea which are presumed to have evolved from deep-sea ancestors.

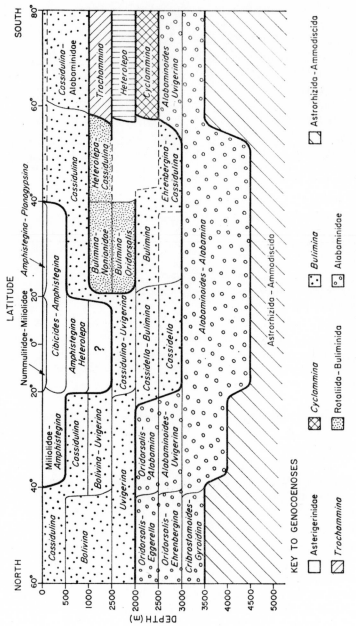

Figure 89 Distribution of foraminiferids in the Pacific Ocean (based on Saidova, 1967b).

Menzies and George (1967) examined endemism among hadal species, and quoted the following data for foraminiferids:

number of hadal species   6000 m   = 128
number of hadal species   <6000 m   = 73
number of species exclusively at depth 6000 m = 55
percentage of hadal endemic species   = 33%

Compared with many other animal groups, the percentage of hadal endemic species of foraminiferids is fairly low. Thus the migration of shallow-water species into the deep sea appears to have been important in the case of foraminiferids.

The mode of reproduction has an important bearing in the derivation of the fauna. Standing crop size is low in the deep sea, so it may be a considerable distance from one specimen to another of the same species, especially in view of the high diversity. There is no reason to expect seasonal control over the time of reproduction because the environment is so stable (although Saidova, 1967a, considers reproduction in the Pacific to take place between August and September; her samples contained up to 80 per cent juvenile living individuals). It is possible, therefore, that sexual reproduction is rare and asexual reproduction is more common. If this is so, the rate of evolution is likely to be very slow. This could account for the great similarity between certain modern deep-sea species and some Palaeozoic forms.

Some generalizations which can be made about deep-sea foraminiferids are:

1. Because of the calcium carbonate compensation depth problem, calcareous shelled forms dominate assemblages in the bathyal zone but are absent from the abyssal and hadal zones (although Bandy and Rodolfo, 1967, have recorded living representatives at a depth greater than 6000 m).
2. Below the calcium carbonate compensation depth, only agglutinated foraminiferids exist.
3. Due to the uniform temperature and salinity conditions, deep-water species are cosmopolitan.
4. The great abundance of benthic foraminiferid tests in deep-sea sediment is related more to the very slow rate of clastic sediment deposition than to the rate of production of tests.
5. The food supply must be organic detritus, bacteria and other non-photosynthesizors and by scavenging or predation.
6. Foraminiferids have been recorded from the deepest parts of the ocean: *Sorosphaera abyssorum* Saidova, 10 687 m in the Kurile–Kamchatka Trench (Menzies and George, 1967).
7. Because of the stability of the environment, diversity may be high (see Sanders, 1969; Slobodkin and Sanders, 1969).

14. Coral Reefs

The term coral reef is applied in a general sense to any community of corals living in shallow water. Typically, the corals construct a framework (often in association with calcareous algae) which in turn provides a great variety of ecological niches for other marine animals and plants.

Modern reefs are best developed where the mean annual water temperatures are 23–25 °C (Wells, 1957). They are therefore restricted to tropical and near-tropical regions. Many corals are controlled by light intensity, and reefs develop only in shallow clear water away from the influence of land-derived sediment. Since the growth activities of the corals result in a topographical feature on the sea floor, this encourages attack from the sea, particularly during storms. Corals are then overturned and broken up, especially if they have already been weakened by worms, sponges and molluscs boring into their corallites, and form a biogenic sediment. Over a period of time, the processes of construction and destruction are closely balanced unless sea-level changes interfere with the state of equilibrium.

Foraminiferids occur in coral reefs in two ways: first as adherent species which assist the construction of the framework, and second as part of the epifauna in the niches within the framework.

Wells (1957, p. 613) lists the constructional forces in order of importance —corals, calcareous algae, foraminiferids and molluscs. However, Myers (1943b, p. 28) states '. . . one of the poorest places to look for foraminifers is the living portion of a coral reef'. When the coral has died, it becomes encrusted with a variety of lime-secreting organisms including the foraminiferids *Homotrema* and *Miniacina*. In reefs on the Yucatán shelf, Logan (1969) recognized a *Gypsina-Lithothamnium* community. This is best developed in the depth zone 30–60 m, although locally it occurs in depths as shallow as 20 m. Both organisms encrust the remains of previous organisms. *Gypsina* develops a crust up to 2 cm thick and 4 cm in diameter. This smooths the originally irregular surfaces of *Lithophyllum* and *Lithoporella* nodules. The ratio of *Gypsina* to *Lithothamnium* is 4:1. There appears otherwise to be little literature on the importance of foraminiferids as frame-builders in coral reefs.

A little more is known of the occurrence of foraminiferids in the reefal epifauna and in the associated reef sediment. However, much more research is needed in this field.

JAVA SEA

In a preliminary note, Myers (1941) noted that where there is an abundant macroflora inside the zone of living corals in the reef areas, species of *Calcarina*, *Peneroplis*, *Elphidium*, *Amphistegina* and *Amphisorus* are common.

BARRIER REEF, AUSTRALIA

On the Heron Island reef, larger foraminiferids live mainly in association with algae on the reef flats. Within this environment they are most abundant in protected waters in the lee of a sand cay and shingle banks, and on the outer part of the central northern reef flat (Jell, Maxwell and McKellar, 1965). In areas of powerful currents or where algal growth is sparse, foraminiferids are absent. Thus they are absent from the sandy moats by the island, sandy areas of the reef flat and the outer pavement of the reef edge. Because the foraminiferids live in protected areas, on death their tests remain in the close vicinity of the living habitat. The species encountered include:

> *Peneroplis planatus* (Fichtel and Moll)
> *Marginopora vertebralis* Quoy and Gaimard
> (particularly associated with *Halimeda*)
> *Amphistegina lessonii* d'Orbigny
> *Calcarina hispida* Brady
> *Baculogypsina sphaerulata* (Parker and Jones)

Rare species include the following:

> *Calcarina calcar* d'Orbigny
> *Alveolinella quoyi* (d'Orbigny)
> *Elphidium craticulatum* (Fichtel and Moll)
> *Operculina bartschi* Cushman

Observations by Fisher (1966) show that the foraminiferids live in association with algae and coral fragments as well as on the coral sand. They are abundant on the following algae: *Colpemenia, Caulerpa, Turbinaria, Halimeda, Padina* and *Chlorodesmis*. At depths of 0–30 cm (at low water) the dominant species are *Calcarina*, and *Baculogypsina sphaerulata* Parker and Jones, with smaller numbers of *Marginopora vertebralis* Quoy and Gaimard and *Peneoplis* spp. These species also live on seaweeds exposed for two to three hours at low water.

Fisher tested exposure effects by putting these species in a petri dish with 5–10 mm of sand and allowing it to dry. After twenty hours all species were still alive.

GILBERT ISLANDS, PACIFIC OCEAN

Todd (1961) described the occurrence of living foraminiferids in twenty-four samples from Onotoa Atoll in the Gilbert Islands. The dominant species of the reef and adjacent areas are (in order of abundance):

> *Calcarina spengleri* (Gmelin)
> *Baculogypsina sphaerulata* (Parker and Jones)
> *Amphistegina madagascariensis* d'Orbigny
> *Marginopora vertebralis* Quoy and Gaimard
> *Cymbaloporetta bradyi* (Cushman)
> *C. squammosa* (d'Orbigny)

In the adjacent lagoon the dominant species are (in order of abundance):

Amphistegina madagascariensis d'Orbigny
Heterostegina suborbicularis d'Orbigny
Marginopora vertebralis Quoy and Gaimard
Spirolina arietina (Batsch)
Cymbaloporetta bradyi (Cushman)
C. squammosa (d'Orbigny)
Elphidium striato-punctatum (Fichtel and Moll)
Schlumbergerina alveoliniformis (Brady)

Minor constituents include four large agglutinated species, miliolids, *Peneroplis ellipticus* d'Orbigny, *Sorites marginalis* (Lamarck), buliminids, discorbids, *Elphidium* spp., and several forms that are normally attached.

Todd also recorded the presence of foraminiferid tests in fish. Most types occur in trigger and surgeon fishes which ingest sediment, algae, coral and small echinoids. However, they are present also in parrot, file and puffer fishes, all of which feed on coral.

BARRIER REEF, BRITISH HONDURAS

The barrier reef runs parallel with the shore and is separated by a lagoon 22–30 km wide. Depths vary from subaerial to 6 m and are controlled by the distribution of actively growing coral heads, together with the presence of sand cays and sinkholes. The reef is cut by a channel 60 m deep and by smaller passages up to 9 m deep. The sediment cover of the reef top between the actively growing coral is skeletal sand (coral, algal, bivalve and other bioclasts) or muddy skeletal sand. Sea-grass stabilizes the sediment surface, particularly in the shallow areas between mangrove and coral-sand cays.

The dominant living species observed on sediment samples from the barrier reef are, after Cebulski (1961, 1962, 1969):

Amphisorus hemprichii Ehrenberg
Amphistegina lessoni d'Orbigny
**Asterigerina carinata* d'Orbigny
**Archaias angulatus* (Fichtel and Moll)
*Borelis pulchra (d'Orbigny)
Clavulina tricarinata d'Orbigny
**Cymbaloporetta squammosa* (d'Orbigny)
Discorbis floridensis Cushman
D. orbicularis (Terquem)
Hauerina bradyi Cushman
H. ornatissima (Karrer)
Peneroplis carinatus d'Orbigny
P. pertusus (Forskål)
P. proteus d'Orbigny
Planorbulina acervalis Brady
P. mediterranensis d'Orbigny

* = confined to the reef tract.

Q. polygona d'Orbigny
*Q. *subpolygona* Parr
 Sorites marginalis (Lamarck)
**Triloculina carinata* d'Orbigny
**T. linneiana* d'Orbigny
**Valvulina oviedoiana* d'Orbigny
 Vertebralina cassis var. *mucronata* d'Orbigny

* = confined to the reef tract.

The standing crop varies from 19 to 2784 per 10 cm², but with the exception of three stations which are greater than 1000 per 10 cm², the majority of samples have values between 100 and 700. The diversity ranges from $\alpha = 4$ to $\alpha = 11$ with an average of 7–8. It is not possible to plot the ratios of the three suborders from the data provided by Cebulski (1961) but the Textulariina can be seen to form only a small component of the assemblages. Cebulski (1969) emphasized the importance of differentiating the living and dead foraminiferids, and commented that it would otherwise have been impossible to separate the different faunas. With the exception of three species, all those diagnostic of the main reef contributed dead tests to the adjacent lagoon environment.

PUERTO RICO

On the Cabo Rojo platform, the reef flat facies, comprising areas of growing coral with sand patches inbetween, has *Amphistegina gibbosa* d'Orbigny as the dominant species. Other forms present include:

 Archaias angulatus (Fichtel and Moll)
 Heterostegina antillarum (d'Orbigny)
 Puteolina proteus (d'Orbigny)
 Asterigerina carinata d'Orbigny
 Sorites marginalis (Lamarck)
 Gypsina discus Goës
 Quinqueloculina spp.

The depth ranges from 4.5 to 70 m, temperature 26–28 °C and salinity 34–36 per mille (Seiglie, 1971).

TRUCIAL COAST, PERSIAN GULF

Small areas of coral growth have been described by Murray (1965a, 1966a, b, c, 1970a, b). The main coral genera are *Acropora* and *Porites*. Patch reefs or banks are composed of small areas of active coral growth separated by areas of dead coral and skeletal sands derived from the destruction of the coral assemblage. Seaweeds, filamentous algae and *Lithothamnium* encrust dead coral. The depths are less than 5 m. The water is hypersaline with salinities in the range 42–45 per mille.

The best living assemblages have been obtained from hairy epiphytic growths on dead coral. The dominant species are *Triloculina* type C to-gether with *Miliolinella* sp., *Quinqueloculina* type B, *Rosalina* sp., and *Verte-bralina striata* d'Orbigny (Murray, 1970b). The α index is 3–4. No data are available for standing crop or biomass. Within a coral bank similarity be-tween samples is high (80.8 per cent, Murray, 1970b). The sediment between actively growing or dead coral heads has only a sparse fauna of *Quinqueloculina* spp. and *Ammonia beccarii* (Linné) (Murray, 1970a).

SUMMARY AND CONCLUSIONS

Although it is commonly accepted that adherent foraminiferids are impor-tant as constructional agents in coral reef communities, there are few specific examples in the literature. The occurrence of foraminiferids as reefal epi-fauna, particularly on dead coral, macroflora and biogenic reefal sediment, is known from several localities. In the Indo-Pacific and Atlantic oceans, larger foraminiferids of the genera *Amphistegina*, *Calcarina*, *Baculogypsina*, *Alveolinella*, *Borelis*, *Marginopora* and *Peneroplis* are commonly present in addition to abundant smaller miliolids and rotaliids. However, in the more extreme environmental conditions of the Persian Gulf (high temperatures, high salinities), these genera are absent with the exception of *Peneroplis*.

Much further investigation of coral reef foraminiferids is needed. The emphasis so far has been on samples of weed and sediment. The exact ecological niches occupied need to be known and the role of foraminiferids as reef-builders needs re-evaluation.

15. Larger Foraminiferids

Larger benthic foraminiferids are an important component of Tethyan Tertiary faunas so that the modern representatives merit special attention. The data for living occurrences are summarized in Table 22. Because so little

Table 22 Summary of the data on modern living representatives of larger foraminiferds (based on Wright and Murray, 1971)

Genus \ Factor	Depth range (m)	Temperature (°C)	Salinity (‰)	Sea-grass	Seaweed	Other flora	Sediment	Coral reef	Energy conditions
Alveolinella									
Davies (1970)	0–6	18–26	39–50				√		high
Gypsina									
Logan (1969)	20–60					√			
Marginopora									
Davies (1970)	0–7.6	18–26		√					
Jell et al. (1965)					√				
Sorites									
Davies (1970)	0–7.6	18–26		√					
Blanc-Vernet (1969)	0–35	22–25*	>37*	√					
Peneroplis									
Murray (1970a)	0–7.5	24–27	40–53	√	√				
Davies (1970)	0–7.6	18–26		√					
Blanc-Vernet (1969)	0–35	22–25*	37*	√					
Jell et al. (1965)						√			
Spirolina									
Davies (1970)	0–7.6	18–26		√					
Blanc-Vernet (1969)	0–35	22–25*	>37*	√					
Amphistegina									
Blanc-Vernet (1969)	5–20	25*	>39*	√					
Seiglie (1970)	9.5–12.5	25–26	34–35				√	√	

* Indicates data from Sverdrup, Johnson and Fleming, 1942.

information is available, the occurrence of dead representatives has also been taken into account in compiling Figure 90.

A conspicuous feature of the distribution pattern is that all occurrences in the oceans are encompassed by the 25 °C surface-water isotherms for the southern and northern summers. In the Mediterranean, a restricted assemblage of larger forms extends into waters with a summer surface temperature of 22 °C. The living occurrences show a total temperature range of 18 to 27 °C (Table 22). The Mediterranean is of particular interest because the larger

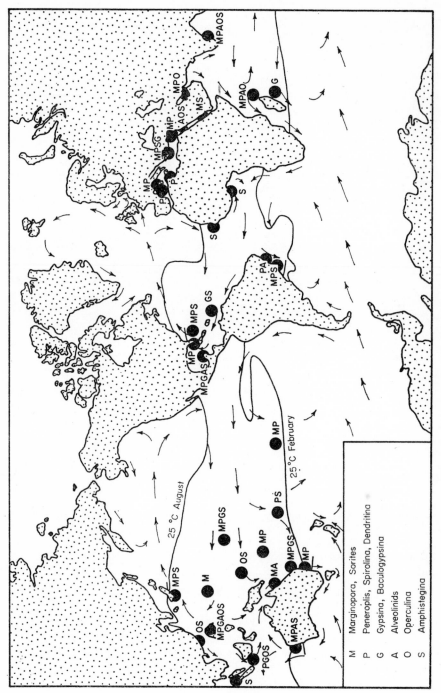

Figure 90 Worldwide distribution of larger benthic foraminiferids (living and dead; isotherms from Sverdrup, Johnson and Fleming, 1942; based on Wright and Murray, 1972)

M   Marginopora, Sorites
P   Peneroplis, Spirolina, Dendritina
G   Gypsina, Baculogypsina
A   Alveolinids
O   Operculina
S   Amphistegina

foraminiferids are present only during the summer months and a minimum summer temperature of 18 °C seems to be necessary for *Peneroplis* and *Sorites* to reproduce. For *Amphistegina*, the critical temperature seems to be close to 25 °C. These genera can colonize areas where the temperature is lower but, as they cannot reproduce, their colonization is temporary.

Due to the nature of modern oceanic circulation, migration from west to east across the oceans is difficult. Although larger foraminiferids are present in the Pacific Islands, they are not found along the Californian and Mexican coasts of America. In this respect the distribution of larger foraminiferids resembles that of modern corals.

All living occurrences are in regions of shallow water (maximum depth 35 m for most genera), where light penetration is good and allows the development of a flora comprising sea-grass, seaweed, calcareous algae and filamentous algae. Field observations show that the foraminiferids live in the protection of the vegetation and the associated microflora probably serves as food. Only *Alveolinella* and *Amphistegina* have been reported as living on sedimentary substrated devoid of a flora. All occurrences are in waters which are fully marine or slightly hypersaline.

However, as can be seen from Table 22, there are no published data on the environment of *Operculina*, the modern analogue of *Nummulites*.

16. Special Features of Tropical Carbonate Environments

In modern marine environments, carbonate sediments are forming in three principal environments, in shallow-water tropical areas, on continental shelves not receiving clastic sediment, and in the deep ocean.

The continental shelf areas of carbonate sediments are areas of slow or non-deposition of clastic sediment. Biogenic material, mainly of molluscan, polyzoan and foraminiferal origin, accumulates slowly. The western English Channel is an example of this type of environment (Murray, 1970c). In the deep oceans, the carbonate sediments result from the accumulation of the remains of planktonic organisms such as coccoliths, pteropods and planktonic foraminiferids. However, it is the third type, the shallow-water tropical sea carbonates, which are the main concern of this chapter. Coral reefs are considered separately in Chapter 14.

TRUCIAL COAST, PERSIAN GULF

The occurrence of living and dead foraminiferids has been described from the shallow shelf of the Persian Gulf (Murray 1966c), the large lagoon, Khor al Bazam (Murray, 1966b), Halat al Bahrani lagoon (Murray, 1966a) and the Abu Dhabi area (Murray, 1965a, 1970a, b). All these studies have shown that foraminiferid assemblages on the bare sediment surface are small and that the main productive areas are associated with the submarine flora.

The Persian Gulf is a land-locked sea which connects with the Gulf of Oman of the Indian Ocean via the straits of Hormuz. Since the surrounding land is arid, there is little contribution of land-derived runoff. The Shatt-al-Arab of Iraq is the main source of fresh water. Evaporation exceeds runoff so that the sea is hypersaline. Data collected by Emery (1956) and Sugden (1963) show a steady increase in salinity with passage west and south into the Gulf. The water is particularly saline along the Trucial Coast. The water temperature is high throughout the year.

General descriptions of the Trucial Coast lagoon and island coastal complex have been given by Evans, Kinsman and Shearman (1964), Evans (1966a, b) and Evans and Bush (1969). Only the Abu Dhabi area has yielded reasonably large assemblages of living foraminiferids, so that this will be considered in some detail.

The Abu Dhabi lagoon lies between the desert land mass of Arabia and the open waters of the Persian Gulf. It is flanked on both sides by low islands. There is a connection with the open Gulf at the seaward end, while at the landward end there are connections around the islands with the adjacent lagoons. The island complex runs from south-west to north-east, perpendicular to the prevailing wind direction (from the north-west). The tidal cycle is irregular, and there is a considerable time delay between high water in the open Gulf and in the inner parts of the lagoon. The tidal range in the open Gulf is 2.1 m maximum compared with approximately 0.6 m in the inner lagoon. Salinities increase from 42 per mille in the open Gulf to 50 per mille in the outer lagoon and 70 per mille in the inner lagoon embayments. The

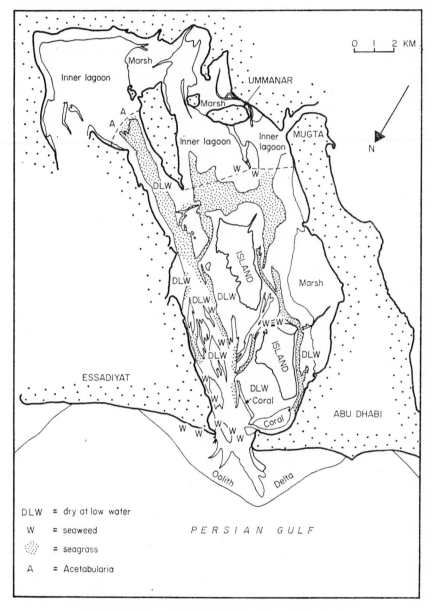

Figure 91 Map of the principal subenvironments and the distribution of the flora in the lagoon adjacent to Abu Dhabi, Trucial Coast.

temperature range from winter to summer in the open Gulf is 20–33 °C, compared with 16–40 °C in the inner lagoon. Thus the environment is one of extremes.

The subenvironments that can be differentiated include the nearshore shelf seaward of the islands, the oolith delta spanning the entrance to the lagoon, the channel, the outer lagoon and the inner lagoon. In addition, there are areas of coral growth near the lagoon entrance and on the nearshore shelf (Figure 91).

Four main plant groups occur in the area, and these are important in the study of the living foraminiferids. Seaweeds grow on rocky subtidal areas, sea-grasses are confined to subtidal soft sediment, mangroves are present in the intertidal zone, and blue-green algae form mats in those areas of intertidal zone protected from severe wave attack. The distribution of the plants is shown in Figure 91.

From samples collected during November and December 1965, Murray (1970a) concluded that living foraminiferids were mainly restricted to sea-

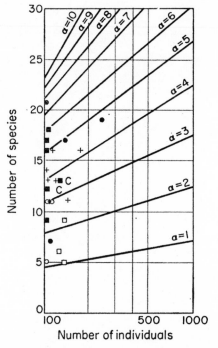

Figure 92 Fisher α diversity indices for the Abu Dhabi region, Trucial Coast (+ oolith delta, ● outer channel, ○ inner channel, c coral bank in lagoon, ■ outer lagoon, □ inner lagoon) (from Murray, 1970b).

weeds. At this time sea-grass was not particularly flourishing. The standing crops were small throughout the area and the same species were present in most subenvironments, with the exception of the inner lagoon, nearshore shelf and frontal reef.

The area was resampled during March 1969. At this time civil engineering works were beginning to cause changes in the lagoon, particularly due to dredging of channels. The local human population had increased five-fold and there was presumably a five-fold increase in the contribution of sewage-derived nutrients. There were significant changes in the distribution of the marine flora, and foraminiferids were abundant. To what extent these changes are related to human activity is difficult to assess.

In 1969 seaweed was sparse but sea-grass (*Halodule*) formed luxuriant submarine meadows over the shallower channels of the outer lagoon. The diversity of the foraminiferid assemblages is low, ranging from α = 1 to α = 7.5 (see Figure 92 and Table 23). Nevertheless, these values conform

Table 23 Summary of diversity, standing crop and biomass in Abu Dhabi lagoon (data from Murray, 1970b)

	Oolith delta	Outer channel	Inner channel	Coral bank	Outer lagoon	Inner lagoon
α index	2.5–5	1.5–7.5	1–3	3–4	2.5–6	1–2.5
Standing crop per 10 cm²	—	1–14	1–7	—	1–38	1–47
Biomass mm³ per 10 cm²	—	0.01–0.42	0.01–0.41	—	0.021–0.58	0.08–4.06

with data from other hypersaline lagoons (see Chapter 3). The triangular plot, Figure 93, shows dominance of Miliolina and occasionally Rotaliina, and a generally low abundance of Textulariina, in all subenvironments.

Standing crop values range from 1 to 47 per 10 cm² but these values exclude seaweed samples (see Table 23). The biomass values range from 0.01 to 4.06 mm³ per 10 cm² of sea floor in the inner lagoon (Table 23). The high values are due to the abundance of the large species *Peneroplis planatus* (Fitchel and Moll). This is the dominant species throughout the area. The associate species include *Peneroplis pertusus* (Forskål), a great variety of *Quinqueloculina* and *Triloculina* species, *Miliolinella* sp., *Vertebralina striata* d'Orbigny, *Elphidium* spp. and *Ammonia beccarii* (Linné) varieties.

Because the foraminiferids live in the protection of the plants, they are in a three-dimensional habitat instead of the essentially two-dimensional habitat of the sediment surface. Thus, to a certain extent, there is small scale geographic isolation between the assemblages on different plants. This is most obvious on the seaweeds. Migration from one plant to another involves a trip either through the water or, more probably, travelling down to the sediment along to the next plant, and ascent to the higher parts of that plant. Nevertheless the assemblages seem to be fairly homogeneous, and comparison of similarities between similar plants (e.g. seaweed/seaweed) and dissimilar plants (seaweed/sea-grass) produces values with a peak at 60 per cent similarity. However, there is very low similarity between plant and sediment assemblages.

It seems most likely that the foraminiferids favour a habitat in the protection of plants because of the high level of organic activity associated with them. The seaweeds support epiphytes among which the foraminiferids

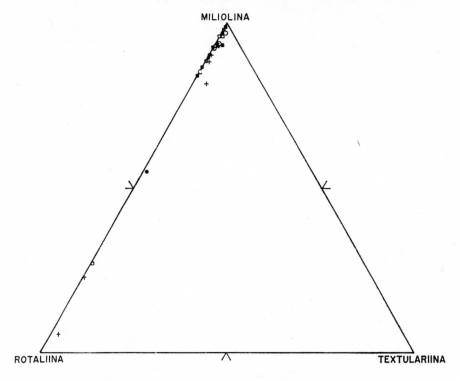

Figure 93 Triangular plot of Abu Dhabi samples
(from Murray, 1970b). Symbols as in Figure 92.

nestle. No doubt food is more readily available there. Likewise, the seagrasses have associated micro-algae which probably serve as food for foraminiferids. By contrast, much of the carbonate sediment surface is either actively moved by currents, e.g. ooliths, or is too clean to provide bacterial food.

SHARK BAY, WESTERN AUSTRALIA

The eastern gulf of Shark Bay has been studied in detail by Davies (1970). The gulf is subdivided into an open northern part and three barred basins in the south. Depths are generally less than 10 m in the southern part. Salinities along the eastern margin, where a bank of sediment is accumulating, vary from 36–56 per mille. The hypersaline conditions result from the land-locked nature of the area, the semi-arid climate, the small land runoff and restricted water movement within the bay. Temperatures are in the range 14–27 °C with local increases to 38 °C.

Three subtidal plant communities are known to be populated by foraminiferids: *Posidonia* sea-grass community, *Cymodocea* sea-grass community and *Posidonia*—algal sea-grass community.

The *Posidonia* community forms dense submarine meadows, particularly in areas of sandy carbonate sediment. In tidal channels it withstands currents of 3.2 km per hour and stabilizes the sediment surface. The plants bear epibionts on their leaves, axes and rhizomes. The principal foraminiferids are:

> *Peneroplis planatus* (Fichtel and Moll)
> *Marginopora vertebralis* Quoy and Gaimard
> *Vertebralina striata* d'Orbigny
> Encrusting nubeculariids
> *Spirillina* sp.
> *Patellina* sp.
> '*Discorbis dimidiatus*' (Parr)

Many of the non-encrusting foraminiferids are embedded in the surface mucus of the sea-grass.

This community extends from low water to a depth of 7.6 m in channels, but below 5.5 m it is reduced in density. In other environments it extends to a depth of 12 m.

The *Cymodocea* community also forms dense submarine thickets. The plants increase in size with depth, and have axes up to 1.2 m long. The epibionts are found at the tips of the leaves. The foraminiferids are the same as those of the *Posidonia* community but *Marginopora vertebralis* Quoy and Gaimard is less abundant. The *Cymodocea* community is best developed on muddy sediments at depths ranging from mean low water to 12 m.

The *Posidonia*-algal community comprises small *Posidonia* plants together with coralline and codiacid algae. The sea-grass cover is only moderately dense and does not completely cover the substrate. Essentially the same foraminiferids are epibiontic on the *Posidonia* as in the *Posidonia* community. Areas of high water exchange and warm temperatures are most suitable for this community. Davies compares it with the *Thalassia* meadows of the Bahamas and Florida Bay.

The carbonate sediments accumulating on the bank contain 45–55 per cent foraminiferids in their skeletal grain component. The majority are of the suborder Miliolina. Little information is available about the occurrence of living individuals on the sediment surface but Davies records *Alveolinella quoyi* (d'Orbigny) in the outer intertidal zone and channel floor. Specimens average 5–8 mm in length and reach up to 12 mm. They do not occur in salinities greater than 50 per mille or in regions of muddy sediment. They are thought to be good indicators of high energy conditions.

In a general discussion of Shark Bay, Logan and Cebulski (1970) record a restricted living assemblage from the hypersaline zone (salinity 56–70 per mille), but it is not clear whether foraminiferids live on the sediment or on weed. Only three species are common:

> *Miliolinella circularis* Bornemann var. *cribrostoma*
>   (Heron-Allen and Earland)
> *Peneroplis planatus* (Fichtel and Moll)
> *Spirolina namelini* (Logan, M. S.)

Less common species include:

Quinqueloculina laevigata d'Orbigny
Quinqueloculina neostriatula (Cushman)
Quinqueloculina seminulum (Linné)
Spirolina acicularis (Batsch)

To the south, in Warnboro Sound, Carrigy (1956) records *Marginopora vertebralis* Quoy and Gaimard living on *Posidonia* sea-grass on shallow sand bars.

THE GREAT BARRIER REEF, AUSTRALIA

Around Heron Island, the larger foraminiferids live in association with algae in protected waters of the reef flats. They are absent from sandy sediments subject to currents and from those areas where algal growth is sparse. Likewise, in the sheltered waters of the lagoon, living individuals are absent because of the restricted occurrence of algae (Jell, Maxwell and McKellar, 1965).

YUCATÁN SHELF

The continental shelf around the Yucatán Peninsula, Mexico, bears a number of submerged rock areas which developed on a land surface that was drowned by the post-glacial rise in sea level. Upon these rocks, organic growth has built up a variety of calcareous structures which have been grouped into three categories (Logan, 1969):

1. submerged organic biostromes,
2. submerged reef banks,
3. emergent reef masses.

The submerged organic biostromes are encrustations of skeletal remains, particularly of nodular calcareous algae of the *Lithophyllum*, *Lithoporella* community, polyzoans and foraminiferids such as *Gypsina plana* (Carter).

The submerged reef banks are composed mainly of corals and the flanks of the supporting banks have an algal nodule—*Gypsina* community.

The emergent reef masses are mainly the products of coral and algal build-up, and there is considerable gross morphological variation. Foraminiferids are most conspicuous on the basal slope where, together with other encrusting organisms, they comprise the *Gypsina–Lithothamnium* community.

Clearly Logan took note mainly of large encrusting foraminiferids, and in particular of *Gypsina plana* (Carter) because of its importance in the *Gypsina–Lithothamnium* community. This develops mainly in the depth range 30–60 m although there is a local development as shallow as 20 m. *Gypsina* forms a smooth encrustation up to 4 cm in diameter and the proportion of foraminiferids to algae is 4:1.

Homotrema rubrum (Lamarck) encrusts corals and coralline algae on the reef flat. It is particularly abundant on the underside of projections down to a depth of about 10 m. At greater depths, *Gypsina* is the dominant encrusting form (Logan *et al.*, 1969). *Amphistegina lessonii* (d'Orbigny) and *Asterigerina carinata* (d'Orbigny) are widely distributed at depths less than 55 m. Other foraminiferids include *Archaias angulatus* (Fichtel and Moll), *Peneroplis* spp., *Sorites* sp. and miliolids but the authors are not specific about their living habitats (Logan *et al.*, 1969, p. 46).

FLORIDA BAY

Moore (1957) examined sediment samples and recorded the presence of living miliolids. However, he was puzzled by the absence of living peneroplids although they form up to 90 per cent of the death assemblage. He did not describe the environment or comment on the presence or absence of marine plants.

CONCLUSIONS

Thus in shallow-water tropical carbonate sediment environments, the foraminiferids are most abundantly associated with submarine plants. This association is not peculiar to such environments but is most conspicuously developed here. Its importance to the geologist lies in the realization that the dead assemblage is derived elsewhere from the sediment surface. In the case of the sea-grass assemblages, dead individuals may drop into the underlying sediment. However, should a storm uproot the grass, the foraminiferids are liable to be transported with the vegetation to different depositional areas. Boch (1970) has used this hypothesis to explain the geographical distribution of plant-dwelling foraminiferids. In the case of the seaweeds, these are growing on rocks which are clearly in areas on non-deposition of sediment. On death, the foraminiferids will be transported to a depositional area. Like the sea-grasses, seaweeds may also be uprooted and transported during storms. Therefore plant-dwelling foraminiferids are more prone to post-mortem transport than are their sediment-dwelling relatives.

17. Communities and Populations

COMMUNITIES

No organism lives by itself. Animals and plants form communities by virtue of their interdependence. The basic communities are sufficiently well known for the associated fauna to be defined by the community name.

Only two foraminiferid communities are known (Thorson, 1957). The Arctic community is characterized by *Rhabdammina cornuta* (Brady) and *Planispirinoides bucculentus* (Brady) (= *Miliolina bucculenta* of Thorson) together with *Asychis* and *Axinopsis*. It occurs on muddy bottoms at depths of 50 to 700 m in fjords in north-east Greenland. The Boreal community is characterized by *Astrorhiza arenaria* Norman, *Saccammina spherica* Sars, *Psammosphaera fusca* Schultze together with *Thyasira*. It occurs in the North Sea at a depth of 100 m. Similar communities with *Astrorhiza limicola* Sandahl are known from shallower waters.

Buchanan and Hedley (1960) made a detailed study of *A. limicola*. Its habitat is on sandy areas of sea floor in the North Sea. The associated fauna, dominated by *Amphiura* and *Nephthys*, is interpreted as a variation of the *Amphiura* community. The ecological niche of *A. limicola* is that of an active predator, crawling at a rate of 1 cm/h for twenty-four hours or so. It then settles down in one spot to develop a pseudopodial net which penetrates the sand to a depth of 2 to 3 mm and extends laterally for 6 or 7 cm. It feeds on live animals such as copepods, amphipods, nematodes and small echinoderms which it traps with its pseudopodia (said to be 'very adhesive' during feeding).

A lot is now known of the habitat of individual foraminiferid species, but very little is known of their ecological niches. Many of them must play an important part in the food chain by feeding on bacteria and possibly on organic detritus. In turn, they are fed upon by predators, particularly deposit feeders and microcarnivores.

In the deep sea, Saidova (1967a) has calculated the biomass to be up to 10 g/m². Boltovskoy and Lena (1969b) quote a comment by Zenkevich that this is sufficient to meet the needs of benthos feeding on foraminiferids. Further, he suggests that the role of benthic foraminiferids can be compared with that of phytoplankton in the surface water.

POPULATIONS

Within the community there are populations of individual species. Certain features are characteristic of populations although they cannot be applied to individuals, e.g. birth rate, death rate, age distribution, density and dispersion.

Birth rate

As applied to foraminiferids, birth rate describes the addition of new individuals (by sexual or asexual reproduction) to the population during a specific unit of time. It seems probable that most foraminiferids reproduce once a year in cool environments and more frequently in warmer environments. The birth rate is thus variable in time; it also varies with environmental conditions and with the size of the population. Maximum birth rate would be expected only under optimum conditions, i.e. optimum environmental conditions and optimum population size.

Death rate

Death rate is the number of individuals dying in a unit period of time. Apart from death through old age, disease or predation, foraminiferids can also 'die' as a result of reproduction. From the geological point of view, the number of tests contributed to the sediment is the most important aspect of death. Reproduction may be the main cause of 'death' and hence of empty shells.

Age distribution

The only easy way of measuring age in foraminiferids is by empirically relating it to size. Within any one study area the relative size should be a reliable guide to age. Where samples have been collected at regular intervals from areas where reproduction is seasonal, individual species populations show rapid size changes. At times of reproduction, the population is composed of small individuals. At times when no reproduction is taking place, most of

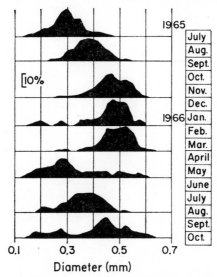

Figure 94 Size distribution of *Elphidium articulatum* (d'Orbigny) (based on Lutze, 1968a, Figure 18).

the individuals are near the average adult size for the species. Where reproduction continues throughout the year, all sizes are represented in the
population, although locally there may be a preponderance of juveniles (due
to recent reproduction) or of adults (about to reproduce). These features
are shown in Figure 94. *Elphidium articulatum* (d'Orbigny) in Bottsand
Lagoon showed a steady increase in size from July to November 1965 related
to the maturing of the population. In March the population reached its
maximum average diameter of 0.5 mm. In April reproduction commenced,
and by May the peak size was 0.3 mm. Reproduction continued until October
1966 but the peak size moved to 0.45 mm, showing that most individuals
were maturing to adults.

Density

Ecologists recognize three main distribution patterns: random, uniform and
clumped. Investigation of various animal groups shows that random and
uniform patterns are rare and clumped distributions are typical of most
species. Among the reasons for clumping, two are perhaps of particular
importance in the case of foraminiferids: micro-environments and reproduction. In variable environments, small, local, differences will favour certain
species, while sexual reproduction requires aggregation of animals such as
foraminiferids.

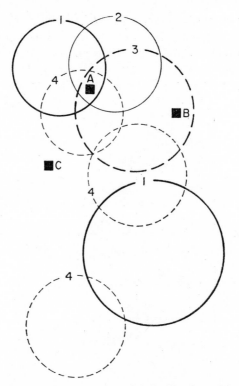

Figure 95 Theoretical model distribution of four
species.

To assist the understanding of the distribution patterns of foraminiferids, two theoretical models are described. A basic assumption is that the local distribution of each species will be a roughly circular area with lowest abundance at the circumference and maximum abundance in the centre. In model 1, Figure 95, four species are represented by circular local distributions which partially overlap. A sample taken at A would contain all four species, while at B only one is present. At C there are no living foraminiferids. In model 2, Figure 96, there are living foraminiferids everywhere. The outline

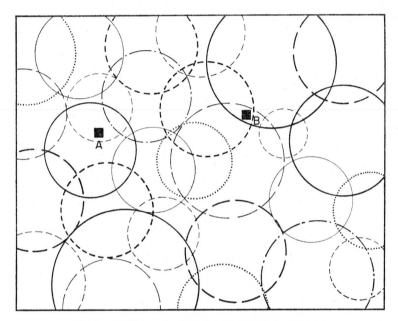

Figure 96 Theoretical model distribution of many species.

square represents the dominant three or four species. Seven rarer species are indicated by the circles. A sample at A will contain the dominant species plus two rarer forms, while at B it will contain the dominant species plus three rarer forms. Note, however, that the rarer species are different at the two points.

How well do these two models fit the observed data? Model 1 has four species, showing a patchy distribution with areas of no standing crop. This closely parallels the situation observed in marshes. Examination of tables of marsh data shows big fluctuations of standing crop size, and the dominant species varies from one sample point to another. Model 2 has a group of dominant species which are present at all sample points, and rarer species which are patchily distributed. The standing crop is fairly uniform as there are no areas lacking living foraminiferids. This pattern is typical of a stable shelf-sea environment.

The two models describe end-points of a gradational series of distribution patterns which range from extremely variable environments (such as tidal marshes) through lagoons and nearshore areas to stable shelf seas.

Parker and Athearn (1959) took duplicate and multiple samples from Poponesset marsh, Massachusetts, and showed that distribution at any one station was fairly uniform although adjacent samples varied considerably.

Lynts (1966) made a special study of the problem. He collected samples from nineteen stations in Buttonwood Sound, Florida Bay, on 14, 17 and 20 August 1962. The standing crop ranged from 4 to 1250 per 18 cm² (corer of 48 mm internal diameter). Environmental differences were thought to be too small to cause changes in the faunal composition, and therefore differences between samples reveal changes in standing crop size. A statistical method was used to show that nine of the standing crops are significantly different. The other ten stations are sufficiently similar to show that the distribution areas for species are at least 30 m². A similar study of Heald Bank in the Gulf of Mexico shows that the distribution areas of individual species are up to at least 2900 m².

Brooks (1967) made a seasonal study of a single station at 20 m depth, in the mouth of the West Passage of Narrangansett Bay. This is a tidal estuary with a bottom salinity of 29–32 per mille and an annual temperature range of 0–21 °C. Preliminary sampling revealed an area rich in living *Ammonia beccarii* (Linné) of limited areal extent. Cores were collected monthly from January to December 1963 and biweekly from July to December. Sub-samples were taken at 1 cm intervals to a depth of 4 cm in each core. Statistical tests using normalization of the data with a transformation of $\log(x + 1)$, where x is the adjusted count, and an analysis of variance showed no significant difference in standing crop size throughout the year. The values ranged between two and six individuals per cubic centimetre. In vertical extent, the standing crop in the upper 3 cm is constant and differs from that between 3 and 4 cm.

Particular attention has been paid to this problem by Buzas. In 1968a he described distributions from twenty-eight samples within one square foot (0.1 m²) of sea floor in Rehoboth Bay, Delaware. He used binomial and negative binomial statistics to determine random and clumped distributions respectively. It was found that only low-density species showed a random distribution, and this was attributed to their being transported into the area. The abundant species showed a clumped distribution because asexual reproduction led to centres of abundance.

A further study of sixteen stations on a grid of 10 m spacing, using five samples from each station and four species, was carried out in Rehoboth Bay (Buzas, 1970). Only *Elphidium* sp. showed a lack of significant difference in standing crop from sample to sample. The other three species, *Ammonia beccarii* (Linné), *Elphidium clavatum* Cushman and *Protelphidium tisburyensis* (Butcher) (= *Elphidium tisburyensis* of Buzas) showed variable standing crop which could not be resolved into a simple pattern of high and low abundance. He concluded that *E. clavatum* had a distribution patch of 100 m², but could not determine the areas for the other two species. Taking all species together they are homogeneous over an area of 1500 m², which means that the total standing crop was not very variable from one station to another.

Lutze (1968b) made a study of three areas: Eckernförder Bucht (near Kiel, Baltic Sea, depth 12 m), Bottsand Lagoon (near Kiel, depth 0.2 m) and Straits of Hormuz (Persian Gulf, depth 65 m). Contoured maps of abundance for Eckernförder Bucht and Bottsand Lagoon show patches of high abundance both for individual species and the total standing crop.

A similar study by Boltovskoy and Lena (1969a) in Puerto Deseado, Argentina, was based on samples collected during September 1967 and January–February 1968. All areas show a patchy distribution. Where the sampling interval is 10 cm, the distance between the centres of the patches is 20 cm. Where the samples are 1 m apart, the distance between the centres of patches is several metres. The authors point out that, over a period of many years, they have observed seventy-six living species at this locality but only forty-four were found in the 1969 study. Clearly the rare species have a patchy distribution both in space and in time.

In regions of submarine vegetation, there is always patchy distribution of the foraminiferids because the plants provide a discontinuous substrate. This relationship was investigated in the lower intertidal and sublittoral areas of North Sea Harbor near New York by Lee et al. (1969). Biweekly samples were collected during the summer of 1966 and 1967. The area was chosen because of its abundant living foraminiferids. Large standing crops occur on various plants, including *Enteromorpha intestinalis*, *Zostera marina* and *Zanichellia palustris*. They are rare on *Fucus* and *Codium*. In early summer, *Enteromorpha* was the principal habitat. The foraminiferids spread to *Zostera*, *Zanechellia*, *Ulva*, *Polysiphonia* and *Ceramium*, and at the end of the summer *Enteromorpha* supported few living individuals.

The living foraminiferids were patchy in their distribution, and areas of high standing crop, termed 'blooms' by the authors, provided most of the individuals. Statistical tests show that the following species are commonly associated in blooms: *Ammonia beccarii* (Linné), *Elphidium incertum* (Williamson), *E. translucens* Natland, *E. clavatum* Cushman, *E. subarcticum* Cushman, *Quinqueloculina seminulum* (Linné), *Miliammina fusca* (Brady), *Rosalina leei* Hedley and Wakefield and *Protelphidium tisburyensis* (Butcher). Negative correlation occurs between *R. leei* and *E. translucens*, *E. subarcticum* and *P. tisburyensis* and *Trochammina inflata* (Montagu) and *Q. seminulum*.

A similar association between foraminiferids and plants was observed by Murray (1970b) in Abu Dhabi Lagoon, Persian Gulf. Seaweeds, seagrasses and algal growths have abundant associated foraminiferids, but the sediment surface supports few live individuals.

Finally, comparison of variations in standing crop and environmental parameters in Choptank River, Maryland, studied over a period of one year, suggests a correlation between standing crop and chlorophyll b (Buzas, 1969). Thus, as one might expect, there is correlation between standing crop size and algal food supply.

The observational data just described fit the suggested distributional models remarkably well, although in practice the distribution patches are often irregular rather than circular. These results have several implications. First, the use of a diversity index will eliminate the problem of patchy distribution of individual species. Second, distribution patterns for individual dominant species will be most reliable for stable environments and least reliable for very variable environments. Third, distribution patterns for rarer species will not usually be meaningful for the small number of samples on which they usually are based. Fourth, all these points must be borne in mind when a sample with an area of 10 cm² is used to represent an area sometimes of many square kilometres.

Dispersal

Little is known of the means of dispersal of foraminiferids. If they relied on their own powers of locomotion, dispersion would be very slow. In shallow water, turbulence causes transport of living and dead tests (Richter, 1965; Schafer and Prakash, 1968). Ice has also been reported as a method of transport (Richter, 1965). Forms living in association with submarine vegetation may be transported when the weeds are uprooted during storms (Boltovskoy and Lena, 1969c; Bock, 1970).

18. Standing Crop, Biomass, Production and Fertility of the Sea Floor

STANDING CROP

The number of individuals on a unit area of sea floor at any moment in time is the standing crop. Values recorded to date range from 0 to at least 4500 per 10 cm². Average values for shelf and marginal marine environments are in the range 50–200 per 10 cm². However, where seasonal changes of standing crop have been recorded, variation at individual sampling points has sometimes been tenfold and is commonly threefold (Haake, 1967; Lutze, 1968a; Boltovskoy, 1964; Boltovskoy and Lena, 1969b; Phleger and Lankford, 1957).

BIOMASS

Normally, biomass is expressed as live weight but this cannot be used for foraminiferids due to their small size. Boltovskoy and Lena (1969b) took 2800 small dry specimens from Puerto Deseado, weighed them, calcined the shells and weighed them again. The dry weight of the protoplasm was 12.4 mg. The biomass of the annual production at this station (1048 per 10 cm²) would be 32.66 mg. Saidova (1967a) determined the wet-weight of protoplasm by multiplying the protoplasm volume by a specific gravity of 1.027. The result for the Kurile–Kamchatka Trench range from intermediate values of 2–3 g/m² to a maximum of 8 g/m².

Volumetric analyses of biomass have been presented by Murray (1968b, 1969, 1970b, c). All available data are summarized in Table 24. The remarkable feature is that the shelf sea values show a very small range (0–0.8 mm³). With the exception of two very high values at Abu Dhabi, the lagoon values fall into the same range. (The foraminiferids have a clumped distribution in Abu Dhabi lagoon and the two high values may be fortuitous). The deep-sea values are much lower than those from elsewhere.

PRODUCTION

Production is the amount or number of organisms resulting from the turn-over of standing crop during a period of time. Productivity is the rate of production. The annual production of foraminiferids can be defined as the number of tests contributed to a unit area of sea floor during the course of one year. Production will depend on four main factors: the initial size of the standing crop, the proportion of individuals which reproduce, the

Table 24 Biomass of living foraminiferids

Locality	Biomass: mm^3 foraminiferids per 10 cm^2 sea floor
Lagoons	
Buzzards Bay[1]	0.20–0.87
Puerto Deseado[2]	(0.31)*
Abu Dhabi[3]	0.004–0.579 + two values 2.777 and 4.006
Shelf seas	
Vineyard Sound[4]	0.007–0.346
Off Long Island[4]	0–0.624
Cape Hatteras[4]	0.003–0.314
English Channel[5]	0–0.219
Bristol Channel[5]	0.132–0.310
Celtic Sea[5]	0.322–0.804
Shelf edge[5]	0.133
Deep sea	
Kurile–kamchatka Trench[6]	0.002–0.077

1. Murray (1968b).
2. Boltovsky and Lena (1969b).
3. Murray (1970b).
4. Murray (1969).
5. Murray (1970c).
6. Saidova (1967a).

Figures in parenthesis have been calculated from weights.
* This value is for the annual production.

frequency of reproduction and the number of new individuals resulting from each reproductive phase. The type of reproduction will not be important except in so far as it affects the number of new individuals.

Information on these factors is limited but the following points arise. Nothing is known of the proportion of individuals which reproduce from a single standing crop. The frequency of reproduction varies from a few weeks to several years. Little is known of the number of progeny resulting from one parent (asexual reproduction) or from two parents (sexual reproduction). However, any given area of sea floor can support a standing crop of a size which is compatible with space and food limitations. Unless there is a variation in food supply, the standing crop size cannot vary too much and an expansion of one species would take place at the expense of another.

Murray (1967b) suggested four basic patterns of relationship between standing crop and production under natural conditions:

Pattern 1 No seasonal changes in the food supply and consequently a uniform standing crop size throughout the year. Reproduction occurs annually for each species, and the number of progeny maturing to adulthood is the same as the number of parents. This is the simplest condition, and the standing crop would be approximately equivalent to the annual production. It is doubtful whether this situation could prevail in shallow seas, but it might possibly do so in the deep sea.

Pattern 2 This would be similar to pattern 1 but reproduction would occur more than once per year in all or some species. The standing crop multiplied by the number of periods of reproduction would give the approximate annual production. This is a more probable situation than pattern 1 and it could be typical of the deep sea.

Pattern 3 Seasonal changes in food supply are matched by an increase in the size of the standing crop. Reproduction occurs annually, but not all species reproduce at the same time. Those species producing juveniles at the beginning of an increase in food supply might achieve a temporary increase in the number of adults. The annual production would differ significantly from the size of the standing crop even though reproduction occurred only once.

Pattern 4 Similar to pattern 3 but having a higher frequency of reproduction. In this case the standing crop would bear little obvious relation to annual production.

Patterns 3 and 4 are probably true of most marginal marine and shelf sea environments.

Thus the relationship between standing crop and production is not simple. In general it can be concluded that where the standing crop is large (more than 1000 per 10 cm²) production will be high even under pattern 1. Where the standing crop is small (less than 10 per 10 cm²), production will be low even under pattern 4. Between these extremes production cannot reliably be assessed from standing crop data.

A possible method of assessing production of single species was put forward by Murray (1967b). Information on the rate of growth of *Elphidium crispum* (Linné) has been replotted in Figure 97(a) (data from Myers, 1942b).

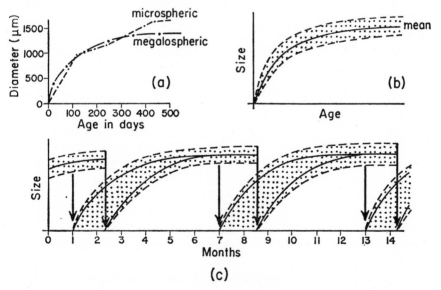

Figure 97(a) Growth curves of *Elphidium crispum* (Linné); (*b*) zone of variability about a mean growth curve; (*c*) effects of periods of reproduction. The short arrows indicate the onset, and the long arrows the termination, of reproduction.

This pattern of growth is probably typical of most foraminiferids. It is possible to superimpose on this curve a zone of variability, as shown in Figure 97(b). In the case of a species which reproduces at a certain season of the year, the living population will change from one composed almost exclusively of adults just prior to the commencement of reproduction, to one composed mainly of juveniles. Reproduction over a long period of time will result in a succession of curves, as shown in Figure 97(c). In this theoretical example, the production of the species could be assessed most reliably by counting the number of adults after the end of production (month 4 in Figure 97(c)). The juveniles which do not reach adolescence or maturity can be disregarded.

For a complete assemblage, production can be based on the dominant species. An example is given below:

Species	Initial standing crop	Mature from reproduction 1	Mature from reproduction 2	Mature from reproduction 3	Total production
A	50	50	—	—	50
B	50	50	50	—	100
C	50	50	50	50	150
D	40	40	40	40	120
E	30	30	30	30	90
					510

The annual production of the five species is 510 individuals. The standing crop remains uniform at 220 individuals. Species A reproduces once, species B twice and species C, D and E three times during the year.

Published results confirm that the method is workable. Boltovskoy (1964) took weekly samples for two years from a creek in Argentina. *Elphidium macellum* (Fichtel and Moll) started to reproduce in November 1961 and juveniles were abundant until January 1962. The adult population was low at the onset of reproduction (138 individuals) but increased to 1703 in January. It then declined to 16 in the following November, which coincided with the next phase of reproduction. Boltovskoy concluded that the life of each individual is not more than one year, that reproduction takes place over a period of six months, and that most reproduction occurs in November (southern hemisphere spring). It is possible to assess production from these figures. The adults from one generation reproduce only once during each reproductive phase. The January figures for both years represent the largest populations of individuals that can be regarded as adults, and as reproduction is annual these figures are close to the true values of annual production. For October 1961 to September 1962, the October standing crop was 113, the peak of mature individuals resulting from reproduction was 1703, and the mature individuals which survived until September 1962 numbered 161. Therefore the annual production is $113 + 1703 - 161 = 1655$.

The decline in the number of individuals between January and October 1962 must represent wastage through disease, competition and natural catastrophe such as burial or predation. This result is typical of pattern 3.

Further data on several small species from Puerto Deseado were provided by Boltovskoy and Lena (1969b). For most of these species juveniles were present throughout the year, and so it was assumed that reproduction took place monthly. These species averaged 86.9 individuals per 10 cm² and rarer species averaged 5.5 per 10 cm². On the basis of monthly reproduction, the annual production was calculated to be

$$5.5 + (86.9 \times 12) = 1048.3 \text{ per 10 cm}^2$$

Daniels (1970) considered the growth period of *Nonionella opima* Cushman in the Limski Channel to be two to three months. There are four to six generations per year. He calculated the annual production at six stations:

Station	Mean standing crop per 10 cm²	Annual production per 10 cm²
39	263	1100–1600
11	168	670–1000
5	260	1100–1600
27	179	700–1100
9	101	400–600
8	98	400–600

FERTILITY OF THE SEA FLOOR

If figures were available for the annual biomass of benthic foraminiferids, they could be used to assess the fertility of the sea floor. At present we have most information on standing crop and some information on biomass at certain moments in time.

Murray (1967b) has already shown that regions of small standing crop must have low production and regions of very large standing crop must have high production. Regions of high production and fertility include the outer edge of the shelf off south California (Uchio, 1960), the eastern Mississippi Delta (Phleger, 1955; Lankford, 1959), and parts of shallow lagoons (Phleger and Ewing, 1962; Phleger, 1956, 1960c). Regions known to be of low fertility and production are the Arctic Ocean (Green, 1960), the northern part of the Gulf of Mexico deeper than 360 m (Walton, 1964a), parts of shallow lagoons (Phleger and Ewing, 1962) and enclosed basins on the Californian shelf (Resig, 1958).

For the remainder of the world, where standing crop values are known to be 50–200 per 10 cm² but the rate of production is unknown, the relative fertility of the sea has yet to be established. The biomass data in Table 24 do not point to any very fertile areas among those so far studied.

19. Living Foraminiferids

This chapter gathers together miscellaneous observations on living foraminiferids, methods of laboratory culture and experiments carried out under culture conditions. The total information is not great but it does have an important bearing on the interpretation of the ecology of the group. However, there is a bias towards shallow-water and intertidal species and virtually a complete lack of information on deeper-water species.

Although the controls which limit reproduction in foraminiferids are of great ecological significance, the actual mode of reproduction is perhaps less so from a geological point of view and is therefore not discussed here. A review has recently been given by Hedley (1964).

CULTURE METHODS

Foraminiferids have been maintained in laboratory cultures by a number of workers starting from the beginning of the nineteenth century. The early methods of collection and culture were simple. They have been described in a review by Myers (1935). In 1954, Arnold described the methods which have tended to be accepted as the standard by subsequent workers. This was followed by a further account in 1966 of special apparatus for culturing small animals. Meanwhile, Lee and a number of co-workers have described advanced methods of maintaining axenic and synxenic cultures.

Living foraminiferids are most readily collected from intertidal areas but, providing boat and sampling facilities are available, can be collected from more or less any sea floor area. In the latter case, one cannot choose the precise area to be sampled unless the samples are collected by a diver. However, on shelf-sea floors there is a reasonable chance of encountering living foraminiferids on the sediment surface. In the intertidal zone, living foraminiferids are restricted in their distribution. The most favourable situations are around the holdfasts of seaweeds or on weeds in rockpools. Algae such as *Corallina* have yielded good assemblages (Hedley, Hurdle and Burdett, 1967). Sometimes rocks encrusted with small algal growths are also good sources. Protected, muddy tidal flats may yield a few living species. The least favourable source area is a sandy beach subject to constant wave attack.

If plants are collected, the foraminiferids can often be dislodged by violently agitating the plants in a bucket of sea water. The residue of debris which falls to the bottom of the bucket can be sieved through a moderately coarse sieve to eliminate the larger plant fragments. Arnold (1954) suggested using a panning technique to separate the smaller organic material from the sand; the latter would gather in the centre of the pan and the former would settle on top. The organic material can be taken off with a pipette.

Once the concentrate has been collected, it should be put in jars for transport unless the laboratory is close at hand. Arnold (1954) suggests that sea water ten to twenty times the volume of concentrate should be added. All collectors have found that it is advisable to cool the concentrate during transport. This can be achieved either by placing it in a precooled vacuum flask or by using large insulated containers. Normally, one should aim to open the jars two or three times a day to allow aeration. This is assisted by shaking the jars to help dissolve the oxygen. Greatest success is achieved if all other animals are excluded (especially worms, molluscs and shrimps) and if the plant debris is kept to a minimum.

On arrival at the laboratory, the concentrate should be put into fresh sea water in shallow, open tanks. Care must be taken to avoid a big temperature difference between the concentrate water and that of the open tanks. Also it is better to keep the tanks by a north-facing window to avoid direct sunlight. Perspex or glass covers are advisable to prevent contamination by dust and to reduce evaporation.

Normally, after a matter of hours or sometimes days, the foraminiferids can be seen on the sediment surface and climbing up the sides of the tank. They can be removed with a paint brush and spatula or with a pipette (taking care to ensure that the aperture is larger than the shell to avoid damage), and placed in freshly prepared culture pots. A variety of containers have been used by different workers. Unless there are larger numbers of foraminiferids, it is advisable to use small, lidded, perspex or glass containers.

It is advisable to observe a few simple precautions when preparing culture dishes. Ideally, the dishes should either be new or have been used only for previous cultures. They should be washed in tap water and rinsed with distilled water. Under no circumstances should detergents be used, and it is inadvisable to wipe the dishes on a cloth that has been washed in detergent. All containers and instruments should be kept away from chemicals which might contaminate the cultures.

The choice of culture medium is important. In many cases the most satisfactory medium is filtered sea water. Such water should be collected well out to sea to avoid land-derived pollution. It should then be filtered to remove organic and inorganic materials in suspension and stored in the dark for one month. Alternatively, artificial sea water can be used; Pantin (1960) lists the following quantities for sea water of 34.5 per mille salinity:

	g	
NaCl	23.427	
KCl	0.729	
$CaCl_2,6H_2O$	2.218	(1.124 anhydrous)
$MgCl_2,6H_2O$	10.702	(5.013 anhydrous)
$Na_2SO_4,10H_2O$	8.967	(3.953 anhydrous)
$NaHCO_3$	0.210	
$NaBr,2H_2O$	0.079	

Sometimes it has been found necessary to enrich the sea water by the addition of nutrients. A common method has been to add earth extract or 'Erdschreiber' according to the methods of Schreiber (1927) and Føyn

(1934). The extract is obtained by autoclaving 1 kg of humus-rich soil in 1 litre of distilled water for 1 hour. After the mixture has stood for several days, the clear liquid is siphoned off. It can be added to sea water together with other chemicals in the ratio:

Sea water	1 litre
Earth extract	50 cm^3
NaNo$_3$	0.10 g
Na$_2$HPO$_4$,12H$_2$O	0.02 g

One major disadvantage is that the quality of earth extract varies according to the type of soil used and its composition is rarely known. Where foraminiferids are to be cultured for ecological purposes, it is probably better to use one of the chemical nutrients described for culturing algae (see Provasoli, McLaughlin and Droop, 1957; Lee, Freudenthal, Muller, Kossoy, Pierce and Grossman, 1963).

Next in importance to providing the right type of culture medium is the supply of food. Normally, cultures of algae are established. Most marine biological laboratories can supply concentrates of individual species of live algae. These can be inoculated into enriched sea water and cultured in advance of the requirements of the foraminiferids. Some authors try to establish algal cultures in the dishes before the foraminiferids are introduced (e.g. Arnold, 1954). Others introduce the food with the foraminiferids. For most efficient feeding, it is often advisable to transfer the foraminiferids to fresh culture medium and food at regular intervals (perhaps twice a week). Two factors are important: an adequate quantity of food must be available (10^3–10^6 cells per 10 cm^3, Lee, McEnery, Pierce and Muller, 1966) and deterioration due to the development of large numbers of bacteria and/or ciliates must be avoided.

Cultures of this kind may be kept at room temperature or at a cooler temperature, depending on the needs of the foraminiferids. Some light is necessary but direct sunlight is to be avoided. Artificial light is commonly used because it can be controlled in intensity and period of exposure.

More specialized requirements of individual species can be determined by experiment. To do this it is necessary to establish synxenic cultures, where all the associated organisms are known, or axenic cultures of the individual species (see Lee, Pierce, Tentchoff and McLaughlin, 1961). The method used is to wash a small group of individuals in a tube of sterile sea water. After ten washings, the individuals are placed in tubes containing 0.2 mg/cm^3 chloroamphenicol, 1 mg/cm^3 dihydrostreptomycin, 0.02 mg/cm^3 polymyxin and 0.5 mg/cm^3 nystatin. After two days, the specimens are again washed ten times in sterile sea water. By this time they should be sterile.

Two different types of elaborate culture apparatus have been described. Freudenthal, Lee and Pierce (1963) established an aquarium system with simulated tides. Arnold (1966) developed an automated culture stand in which there is continual renewal of the culture medium.

FOOD AND FEEDING

Many authors have recorded observations on food and feeding in foraminiferids. Often this has been in the form of incidental information in a paper concerning life history or some other aspect of foraminiferid activity. Sandon (1932) summarized the information available up to that date. Some of the observations are recorded in the following.

It is only in very recent years that experiments have been carried out not only to determine what foraminiferids eat, but also to see what their food requirements are for continued growth and reproduction. Even now, very little is known of the food requirements.

Most of the species that have been used in experimental studies are either intertidal or very shallow subtidal forms. All are from the photic zone and all have been found to feed on live algae. However, at depths beneath the photic zone foraminiferids must feed by predation, by scavenging or on bacteria.

Those foraminiferids that feed on algae normally gather them up with their pseudopodia and pass them towards the aperture. Some forms, like the miliolids, gather the food in a mass about the aperture. Others, such as *Elphidium crispum* (Linné) form distinct feeding cysts around themselves (Jepps, 1942, 1956; Murray, 1963). *Bathysiphon filiformis* Sars accumulates the waste products from feeding at one end of the test. This is periodically discarded. Sheehan and Banner (1972) have suggested that the free-standing pseudopodial nets would be efficient filters of food in suspension.

One species appears to be a scavenger; Hedley (1958) recorded that *Haliphysema tumanowiczii* Bowerbank fed on diatoms, pieces of seaweed and of crustaceans. However, he believed diatoms to be the chief food source. Hofker (1931) recorded foraminiferids feeding on detritus in the Zuidersee.

There have been a number of descriptions of carnivorous feeding. Christiansen (1964) described how the pseudopodia of a young *Spiculosiphon radiata* Christiansen came into contact with a small *Nonion* and had attached themselves to it within twenty-two minutes. Five days later, the *Nonion* was found to be free of protoplasm when tested by staining. A great variety of calcareous species were killed for food, but no agglutinated species. *Nonion* formed 45 per cent of the food, and experiments showed that most were eaten within six days of coming into contact with *Spiculosiphon radiata*.

The pseudopodia of *Astrorhiza limicola* Sandahl were reported to have two physiological conditions by Buchanan and Hedley (1960). In one condition the pseudopodia would not adhere to particles, while in the other condition they adhered to any object they touched. The latter pseudopodia are used during feeding, and the animals caught during experiments include cumaceans, caprellids 2–3 cm long, small sea urchins, various small crustaceans, and *Artemia*. Once caught the animals struggled until they died from exhaustion. No evidence was found that the foraminiferids could kill the prey with toxins. Digestion is believed to be extracellular. Buchanan and Hedley suggest that as the animal is non-selective in the material used for test construction, it is also a non-selective feeder which eats any of the interstitial microfauna it can catch.

However, Heron-Allen (1915) recorded an instance of a copepod falling to the bottom of a tank after contact with miliolid pseudopodia, and suggested that it had been stunned or terrified by its experience. *Rosalina carnivora*, described by Todd (1965), lives attached to shells of *Lima* and may feed on the mantle of the bivalve.

In some of the earliest experiments on feeding, Bradshaw (1955) established that the small foraminiferid *Rotaliella heterocaryotica* Grell could feed both on living *Nitzchia* and dead *Chlamydomonas*, whereas a related species, *Rotaliella* sp., would feed only on living algae. However, more dead *Chlamydomonas* was required to produce a growth curve comparable with that of the population of *Rotaliella heterocaryotica* fed on living food. It was suggested that the concentration of food was important, and that too much may be as bad as too little. Excess food is attacked by bacteria; it can cause chemical side-effects and it can choke the foraminiferid cultures. However, all these adverse effects may be peculiar to the conditions experienced in cultures and need not be typical of conditions in the open sea.

Bradshaw (1961) ran parallel experiments on *Ammonia beccarii tepida* (Cushman). The food was killed *Dunaliella*. In one set of cultures different quantities of food were added: 0, 55 000, 110 000, 220 000, 440 000 and 880 000 cells per culture. In the parallel cultures, the same quantities of food were used with the addition of 1400 units of penicillin G and 1.5 mg streptomycin to eliminate bacteria. The specimens were fed twice-weekly. The cultures treated with penicillin showed a slightly faster rate of growth. Otherwise, both sets of cultures showed that the rate of growth increased with additional food. Food concentrations less than 220 000 cells per culture did not allow any growth or reproduction in the foraminiferids so that this value must be regarded as a minimum. Smaller quantities of food can keep the animals alive but starvation causes a gradual reduction in the protoplasm and the last-formed chambers become empty.

Some simple experiments on feeding in *Elphidium crispum* (Linné) were described by Murray (1963). This animal gathers the food about itself as a feeding cyst which is discarded in due course (Jepps, 1942). *Phaeodactylum tricornutum* Bohlin was used as food. A culture of 10 *Elphidium crispum* in sea water produced 6–11 feeding cysts daily, and averaged 7.5. A culture of 5 individuals in artificial sea water (based on Pantin, 1960) averaged 4.6 feeding cysts per day (= 9.2 for 10 individuals). Initial experiments using dead food, *Phaeodactylum* killed by ultraviolet radiation or by heating, led to a cessation of feeding. However, subsequent experiments showed that dead food was accepted but that the rate of feeding cyst production fell. Thus, dead *Phaeodactylum* appears to be less acceptable than living.

When *Elphidium crispum* was cultured in dishes having a substrate of inorganic particles having the same size range as the food (kaolin, graphite and calcium carbonate), it formed feeding cysts from it within one day. The specimens with graphite 'feeding cysts' were fixed in Bouin's fluid and thin-sectioned. No graphite had entered the chambers confirming Jepps' observation that digestion is carried on outside the test. As the protoplasm had been stained with rose Bengal, it could be seen that the feeding cyst was held together by some other means, perhaps by the secretion of mucus (Jepps,

1942). Discarded feeding cysts retain their form for several days. In other specimens the inorganic feeding cysts were discarded after a few days, showing that the particles had not just stuck to the animal. This experiment suggests that *Elphidium crispum* selects its food on the basis of size. Similar experiments on *Quinqueloculina* spp. show that it, too, picks up inorganic particles having the same size as its food.

Lee, Pierce, Tentchoff and McLaughlin (1961) found that in monoxenic (one associate) or limited agnotobiotic (bacterized) cultures, no individual alga was able to support continuous growth and reproduction of the foraminiferids. Other cultures of *Bolivina* sp. were fed with two algal species. Half the cultures were kept in total darkness and half in subdued light. After six weeks the length of the individuals was measured. There was less growth in those kept in total darkness. In both sets of cultures, those fed with *Nitzchia* sp. had not grown and some had suffered the loss of the terminal chamber, those fed with *Nitzchia acicularis* had grown markedly, and in the other case only slight growth had occurred (*Rhodomonas–Isochrysis* food).

In cultures of the flagellate *Dunaliella parva*, the random motion of the individuals stops if an *Ammonia* is introduced; instead, they swim towards the foraminiferid. The same phenomenon is observed with *Bolivina*, *Quinqueloculina* and *Miliammina fusca* (Brady). The flagellates become trapped by the pseudopodia and passed towards the aperture. The apparent attraction for the flagellates has been termed the 'Circean effect' (Lee and Freudenthal, 1964).

Monoxenic cultures of *Allogromia* sp. NF in two artificial media (ASP$_1$ and ASP$_7$, Lee and Pierce, 1963) produced bigger populations than those in sea water. However, in a brief discussion of prey and predator relationships in littoral foraminiferids, Lee, McEnery, Pierce and Muller (1966) noted that feeding is erratic if the food cells are less than 10^3 organisms per 10 cm^3 culture. Feeding is directly proportional to concentration in the 10^3–10^6 range. Interaction was observed, for *Platymonas subcordiformis* encouraged ingestion of *Nitzchia acicularis*, *Cylindrotheca closterium* and *Dunaliella salina*. Conversely, feeding was reduced in the presence of *Saccharomyces cerevisiae* and species of marine yeasts.

With respect to predation by two species of *Allogromia*, for some foods the presence of the two species together had no effect. However, *Allogromia* sp. (NF) was less efficient than *A. laticollaris* in feeding on *Phaeodactulum tricornutum*, *Chlorococcum* sp. or *Platymonas subcordiformis*. The explanation is that *Allogromia* sp. (NF) feeds on bacteria as well as on algae. Consequently it reduces the bacteria and produces culture conditions favourable to *A. laticollaris*. If too many *Allogromia* sp. (NF) are present, they cannot feed on bacteria alone so they compete with *A. laticollaris* for algae.

The importance of a variety of food, including bacteria, was demonstrated by experiments on *Spiroloculina hyalina* (Lee and Muller, 1967). Both gnotobiotic bacterized and gnotobiotic bacteria-free cultures were established. In bacterized cultures with single or paired algal associates, reproduction took three to four times longer than the controls. The addition of more algal species shortened the generation time. Bacteria-free cultures were successful also but the generation time was six months. Addition of two species of bacteria shortened the time to four months.

The most sophisticated feeding experiments have been those using radioactive tracers. Detailed methods have been described by Lee, McEnery, Pierce, Freudenthal and Muller (1966). They tested more than fifty potential food organisms on a variety of foraminiferids from shallow-water environments. Yeasts, blue-green algae and dinoflagellates were not generally accepted as food. The most popular were diatoms (*Phaeodactylum tricornutum, Nitzchia acicularis, Cylindrotheca closterium*), a chlamydomonad (*Dunaliella parva* and a chrysomonad (*Monochrysis lutheri*). The age of the foraminiferid with respect to the life cycle was important. Young *A. laticollaris* (150–200μm) ate much more food than older individuals (350–500 μm); sometimes the difference was 200 per cent. As with previous experiments, feeding was proportional to the concentration of food. The age of the food was also of importance for some species, such as *Ammonia beccarii*, which had a clear preference for food less than ten days old.

Further experiments (Muller and Lee, 1969) using bacterized and bacteria-free gnotobiotic cultures, confirmed that the presence of bacteria is necessary for reproduction to take place. It is thought that they provide some nutritional need which algae alone cannot fulfil.

Lee, McEnery, Pierce, Freudenthal and Muller (1966) suggest that many littoral foraminiferids are bloom feeders: 'When low concentrations of mature food organisms are present, foraminifera eat and reproduce slowly; when large quantities of the appropriate food organisms are present in the form of a vigorously reproducing bloom, the foraminifera may exploit this abundance'.

Thus foraminiferids living in the photic zone are selective in what they eat, and require a mixed diet of algae and bacteria for optimum growth and reproduction. Nothing is yet known of the food requirements of foraminiferids living below the photic zone.

CALCIUM UPTAKE

McEnery and Lee (1970) used ^{45}Ca and ^{90}Sr to trace mineral cycling in *Rosalina leei* Hedley and Wakefield and *Spiroloculina hyalina* Schultze. They also introduced ^{32}P, ^{14}C and ^{35}S in the range of $1 \times 10^2 - 1 \times 10^7$ dpm/cm^3 of medium (dpm = disintegrations per minute). The same food algae were supplied to each experimental flask. Results of short- and long-term experiments show that ^{90}Sr, ^{45}Ca, ^{35}S and ^{32}P are good indices for growth in the two calcareous foraminiferids when used in a concentration of $1 \times 10^4 - 1 \times 10^5$ dpm/cm^3 of medium. Once incorporated in the test, calcium and strontium are not exchanged with the environmental medium unless acidic conditions prevail. Decalcification was carried out in sea water with 0.01 per cent Na$_2$ EDTA at pH 8.1. This did not harm the specimens and they were able to recalcify their tests when returned to the control medium.

POLLUTION

A problem affecting modern foraminiferids is pollution. Lee and Marcellina (1967) briefly reported the results of experiments using pollutants on *Allogromia laticollaris*. The presence of pollutants temporarily or permanently inhibited reproduction. However, the ability to feed was not impaired, although the foraminiferids were found to be underfed. This was thought to be due to the adverse effects of the pollutants on the food rather than on the foraminiferids.

Field studies of pollution are described on pages 50, 103–5.

ILLUMINATION

It has been stated that 'most foraminifera are positively phototropic and negatively geotropic' (Myers, 1943a, p. 453). At times of reproduction, foraminiferids move higher on objects on the sea floor to aid the dispersal of gametes or progeny. During the autumn Myers observed a reverse trend, the foraminiferids retreating to crevices. However, whether these movements are really correlated with illumination is open to question. An experiment by Murray (1963), using a positively phototropic food alga, *Tetraselmis suecica* and *Elphidium crispum*, resulted in the foraminiferids not following the food to the light side of the culture dish. In a similar experiment using *Rosalina globularis* d'Orbigny, Sliter (1965) observed that the foraminiferids followed the diatoms to the light side of the dish but he attributed this to a secondary response.

Sheehan and Banner (1972) experimented with *Elphidium incertum* (Williamson) from floating weed. These were placed in a container with a point source of light on one side. After three days, 80 per cent of the specimens had climbed up to the light source. In a second experiment, individuals from muddy sediment were kept in the dark for four days. At the end of this time, 20 per cent had climbed up the walls, showing negative geotropism. However, other experiments showed no consistent pattern of behaviour.

COLOUR OF THE PROTOPLASM: SYMBIOSIS

The majority of shallow-water benthic foraminiferids have coloured protoplasm: green, brown, purple, orange and red. In general, this colouration has been attributed to three origins—the presence of symbiotic algae, the colour of the food and pigments.

Many authors have drawn attention to the presence of symbiotic unicellular algae in the protoplasm since Winter (1907) first recorded them in *Peneroplis*. Boltovskoy (1963a) has summarized the available records. He notes that algae are present even in forms with small apertures. Cushman (1922) reported that the algae in *Iridia* were carried outside the test by the pseudopodia when the animal was active. During rest periods the algae were stored within the test. Haynes (1965) suggested that the foraminiferid tests provide a glasshouse for the algae, particularly when the wall structure is radial. In species from the Dovey estuary, he found algae plastering the inner surface of the chamber wall. This was interpreted as an advantage for interchange between the protoplasm and the sea water outside via the wall pores. Further, Haynes believes that the algae lead to modifications such as complex canals and to shapes which permit greater surface exposure to light, e.g. planispiral compressed tests. He quotes a number of examples from the fossil record to support this claim. The porcellaneous wall, as seen in the peneroplids, he interprets as an adaptation to well-lit waters where scattering of the light by the wall crystals would reduce the ultraviolet light.

Hedley (1964) discounted claims of zooxanthellae symbionts in foraminiferids, with the exception of those observed in *Orbitolites duplex* Carpenter by Doyle and Doyle (1940). However, Lee and Zucker (1969) have cultured the algal symbionts of *Archaias* from Florida. The algae are important in controlling the uptake of ^{45}Ca; maximum uptake occurs with maximum light intensity. They enable the foraminiferid to take up calcium more efficiently and therefore they increase the rate of test calcification. Infection takes place during feeding. The protective sheath of the algae prevents their destruction during feeding, even when they are ingested by other species such as *Rosalina leei* Hedley and Wakefield. Thus it seems clear that some shallow-water foraminiferids do have a symbiotic relationship with algae.

Nevertheless, some foraminiferids clearly derive their colour from their food. During the course of experiments with *Elphidium crispum* (Linné), individuals fed with brown coloured *Phaeodactylum tricornutum* Bohlin had brown protoplasm. One culture was fed with green alga *Tetraselmis suecica*; after eight days' feeding the foraminiferids had turned yellow, and after nineteen days they were green (Murray, 1963).

PARASITISM

The occurrence of egg cases of turbellarians attached to foraminiferid tests has been reported by Ehrenberg (1839), Schultze (1854), Jepps (1942), Boltovskoy (1963b) and le Campion (1970). Two different interpretations have been made: Jepps and le Campion consider the turbellarians to be parasitic; Boltovskoy believes them to be epibionts.

Lister (1895) noted the presence of an unidentified parasite in *Orbitolites complanatus* Lamarck from Tonga. Le Calvez (1940) recorded an amoeba, *Wahlkampfia discorbini* le Calvez, parasitic on *Discorbis mediterranensis* (d'Orbigny). Myers (1943a) found developmental stages of *Trophosphaera planorbulinae* (le Calvez) in 3 to 5 per cent of *Elphidium crispum* (Linné) collected from the subtidal zone in Plymouth Sound in the spring. At the same time, nematode worms were present in the tests.

Two instances are known of one foraminiferid being parasitic on another (le Calvez, 1947). *Fissurina marginata* (Montagu) (= *Entosolenia marginata* of le Calvez) goes into the pseudopodial reticulum of *Discorbis villardeboanus* (d'Orbigny) and takes out granules. *Planorbulinopsis parasita* Banner is an endoparasite of *Alveolinella quoii* d'Orbigny in the Coral Sea. It excavates holes 0.4 mm in diameter and up to 0.1 mm deep in the test of *A. quoii* (Banner, 1971).

Todd (1965) described *Rosalina carnivora* Todd living in excavated areas of the shells of *Lima* (*Acesta*) *angolensis*. The foraminiferid sometimes corrodes through the shell to feed on the mantle. It is probably a carnivore rather than a parasite.

FUNCTION OF THE TEST

In an excellent review of the subject Marzalek, Wright and Hay (1969) suggest the following possible functions of the test of *Iridia*:

1. to protect the foraminiferid against predation;
2. to provide shelter from unfavourable physical or chemical conditions;
3. to serve as a receptacle for excreted matter;
4. to assist the processes of reproduction;
5. to control the buoyancy of the organism.

The only possibility they favour is the buoyancy function. After feeding, a foraminiferid may develop reserves of light-weight lipids and fats. These might produce buoyancy which the pseudopodial attachments to the sea floor would be unable to counteract. Therefore, addition of a relatively dense shell would be advantageous.

In the case of *Bathysiphon*, the distinctive tubular form might be for protection against predators or from unfavourable physical or chemical conditions, especially if the animal is able to plug the ends of the tube. Likewise, chambered tests would be effective barriers to osmotic changes in the ambient medium. The choice of *Quinqueloculina* as an example resisting osmotic changes is unfortunate since experimental work reveals this to be untrue (Murray, 1968b).

Rotaliids have perforate walls and some have well-developed canal systems. The pores are not solid, as suggested by Angell (1967). They are traversed by organic pore plates through which minute pseudopodia emerge (Hansen, 1972). The function of the latter is probably to keep the surface of the test clean and free from epibionts.

Marszalek *et al.* (1969) suggest that *Elphidium* may use its test as a greenhouse for culturing symbiotic algae while the protoplasm would occupy the canal system. This arises from a misinterpretation of the structure of the canal system. Sections of *Elphidium crispum* (Linné) prepared by the author show that the protoplasm occupies the chambers but not the canal system. The latter appears to have no direct connection with the inside of the chambers. All the openings are to the outside of the test.

Haynes (1965) has suggested correlation of wall structure, habitat and the presence of symbiotic algae in foraminiferids, but this has yet to be proved.

One feature that is generally overlooked is that unchambered tests could result from continuous growth whereas chambered tests are clearly the product of discontinuous growth (Glaessner, 1963). Further, the changes resulting from chambering and coiling add much greater strength to the test. Added to this are the observed differences of shape associated with environmental differences (discussed under 'relationship with the substrate, pages 217–18). Clearly, the function in some cases at least is protection from physical and biological adversities.

MOVEMENT

Movement is accomplished by the pseudopodia although very little is known of the mechanism by which these function (see Hedley, 1964; Sheehan and Banner, 1972). Normally, an active individual raises its test upright after a dense pseudopodial net has spread in all directions over the substrate. Movement in a given direction is discontinuous; pseudopodia are extended in the 'forward' direction and then the shell is drawn up, rather as with a snail. When at rest, the shell normally lies on the substrate. Descriptions of the pseudopodia during movement have been given by Sheehan and Banner (1972).

It is commonly observed in freshly collected material, that the foraminiferids climb up on to the sediment surface and they also climb up the sides of culture vessels. This behaviour can readily be associated with the need to regain the sediment surface after burial in the sediment during storm reworking of the sea floor.

Some authors believe that pseudopodia exist in two physiological states, adhesive and non-adhesive (Buchanan and Hedley, 1960). Jepps (1942) had observed that pseudopodia are bathed in mucus, and had discussed the possibility of pseudopodial 'end-knobs' being sticky. Sometimes a foraminiferid will move through food without picking it up, whereas at another time it would adhere and be taken up by the pseudopodia.

An unusual method of locomotion was recorded by Jepps (1942). She observed *Elphidium crispum* (Linné) to attach itself to the surface skin of the water by its pseudopodia so that it hung suspended. This has been observed in cultures of the same species by the author. Richter (1965) has proposed the same mechanism as a means of transport of living foraminiferids in the Jade Sea off Germany.

Experiments on movement carried out by Arnold (1953) suggested that the rate of movement of *Allogromia laticollaris* was inversely proportional to the age of the laboratory cultures. In a well-balanced culture there is less movement than in a newly established culture. The maximum rate of movement observed was 10 mm/h but more average values were 2.5–2.9 mm/h. Arnold considered that his results showed strong dispersal, particularly of juveniles moving away from the parent tests. However, in older cultures the animals became evenly distributed. Sliter (1965) recorded a maximum value of 12.5 mm in 3 hours for *Rosalina globularis* d'Orbigny. He noted a strong tendency to disperse where food organisms are sparse. As the quantity of food increases, mobility decreases.

In an experiment using corn-starch, Arnold tried to discover whether *Allogromia* would be attracted to the food. No attraction was found; the animals came across the starch by accident and then moved elsewhere. No response to light was observed, and it was concluded that movement proceeded as much in darkness as under daylight conditions.

Murray (1963) placed 10 *Elphidium crispum* (Linné) along a centrally placed line at 2 mm intervals in a square culture tank. During the period of observation, 5 individuals were present to each side of the line, suggesting no preferred direction of movement. However, in another experiment in which half the culture tank was lined with kaolin, 5 individuals were placed in each half, and at the end of 14 days' observation 3 remained in the clay half while 7 were in the clear half. Presumably, once an individual had escaped from the clay substrate it was not keen to return and therefore movement was not entirely random.

Specimens of *Ammonia beccarii* (Linné) observed by the author show a number of interesting features. Not uncommonly, the last chamber is apparently devoid of protoplasm and yet pseudopodia emerge from its aperture. Because of the position of the aperture and the trochospiral nature of the test, the pseudopodia arise from the ventral (involute) surface. When the animal moves along, the shell is held erect (i.e. with dorsal and ventral sides pointing laterally) and the pseudopodial net arising from the aperture. In the resting position, the animal can be dorsal or ventral side uppermost. If the ventral (involute) side is uppermost, problems arise when the animal wishes to move. The pseudopodia are extended from the aperture over the side of the test to the substrate. When sufficient anchorage has been achieved, they contract and pull the test upright.

Thus, foraminiferids move by means of their apertural pseudopodia. They are erratic in speed and direction.

RELATIONSHIP WITH THE SUBSTRATE

The main substrate types are unconsolidated sediment and firm surfaces such as rocks, seaweeds or other plants.

It is generally assumed that foraminiferids live at or near the sediment surface. (Their penetration beneath the surface is considered in the next section.)

Nyholm (1957) recorded observations on the orientation of the test and the use of the pseudopodia by simple foraminiferids living on soft sediment surfaces, particularly those rich in organic detritus. *Gromia* has more rigid pseudopodia which penetrate the sediment to a greater depth than those of *Hippocrepinella* and *Tinogullmia*. In the latter, the pseudopodia are slender and fragile and consequently do not penetrate deeply into the sediment.

Many of the forms with single apertures investigated by Nyholm hold their test perpendicular to the substrate, aperture downwards. Other forms with two apertures were found to rest horizontally in the sediment, and some (*Phainogullmia* and *Nemogullmia*) gathered much detritus between their pseudopodia. The binding power of such foraminiferids obviously could be important in limiting the erosional effects of turbulence on the sea floor.

Some foraminiferids with a single aperture have weak pseudopodia and therefore remain horizontal in the sediment. *Bathysiphon* stands erect and is able to collect detritus from the sediment at the lower aperture and from the water at the upper aperture. Because the tubes are flexible, their orientation changes in response to changing currents.

Astrorhiza limicola Sandahl leaves a furrow in the sand when it moves. Buchanan and Hedley (1960) found that movement may occur for twenty-four hours, then the animal settles and produces a pseudopodial system which penetrates the sediment to a depth of 2–3 mm, and laterally 6–7 cm.

Sliter (1965) recorded small differences of mobility related to sediment type. *Rosalina globularis* d'Orbigny moves as easily on fine sand as on the smooth floor of culture pots. Those on coarse sand move very little.

For most species, nothing is known of their relationship with the sedimentary substrate and this is clearly an important field for research.

In shallow waters where there are seaweeds or rocky substrates, foraminiferids seek protection from surf in one of four ways:

1. by cementing themselves to the substrate, e.g. *Acervulina, Marginopora*;
2. by clinging to the substrate by means of their pseudopodia, e.g. *Spirillina, Patellina, Rosalina*;
3. by seeking the protection of crevices, usually near the holdfast, or in dense weed growths, e.g. *Calcarina*;
4. by hiding in dense epiphytic growths of *Jania* and similar algae, e.g. *Peneroplis*.

In the deeper water of the Java Sea, where there are dense seaweed growths less subject to surf, Myers (1942a) records *Elphidium* and *Amphistegina* with thick biconvex tests. *Tinoporus* attaches itself by pseudopodia from the tips of its spines (Myers, 1943b). *Baculogypsina* behaves in a similar fashion and is found at depths of 20–60 m.

In quieter tropical and subtropical waters where sea-grass grows, discoid species such as *Sorites* and *Amphisorus* cling or cement themselves to the blades. Other free-living species, such as miliolids and peneroplids, crawl over the plants.

Myers (1943b) concluded that larger foraminiferids, with discoidal, planispiral or fusiform tests, are restricted to the sublittoral zone in tropical seas, particularly those with marine plants growing adjacent to reefs with corals, calcareous algae and polyzoans, and on firm sandy or sandy mud substrates.

In the Gullmar Fjord, Nyholm (1961), found that ascidians living in the 40–50 m depth zone had an epizoa of *Smittina* colonies, *Serpula vermicularis*, *Hydroides norvegica*, *Verruca stroemi* etc. on one side, and foraminiferids, notably *Cibicides lobatulus* (Walker and Jacobs), on the other. These foraminiferids live with their flat side held close to the substrate by the umbilical pseudopodia. Nyholm sectioned *Cibicides* on ascidians and established that the pseudopodia do not penetrate the surface.

DEPTH OF LIFE IN THE SEDIMENT

Normally, samples for ecological study are collected from the top centimetre of sediment and this presupposes that foraminiferids are confined to this depth zone. Myers (1943a) concluded that *Elphidium crispum* (Linné) living on sand in Plymouth Sound, England, could withstand burial to a depth of 1 cm. He suggested that foraminiferids could survive burial equal to five to seven times the diameter of the test. Deeper burial would result in death. Furthermore, this species becomes dormant in the winter months, and if buried would make no attempt to regain the sediment surface. Myers concluded that 80 per cent of the adult individuals in the subtidal zone were killed by burial resulting from turbulence and currents. Mortality on this scale could be a limiting factor for a species.

Green (1960) stained samples from Arctic Ocean cores with rose Bengal. He found that even at a depth of 20 cm below the sediment surface some foraminiferids were stained, and he advised caution in using this staining method.

Richter (1964b) found in the Jade Bay of Germany that the foraminiferids lived at different depth zones in the intertidal sediments. *Elphidium articulatum* (d'Orbigny) (= *E. excavatum* of Richter) inhabits the surface; it never voluntarily penetrates into the sediment, and if it should get buried it rapidly tries to regain the surface. Normally, it leaves small trails or tracks on soft sediment surfaces. *Protelphidium anglicum* Murray (= *Nonion depressulus* of Richter) rarely lives on the sediment surface; usually it occurs immediately under the surface. Cores from the tidal flats showed *Elphidium articulatum* exclusively on the surface of the sediment, *E. excavatum* (Terquem) (= *E. selseyense* of Richter) at a depth of 0.5–6 cm and *P. anglicum* from the surface to about 3 cm. This partly accounts for the ability of the latter two species to live in areas of strong currents with associated sediment movement, and the confinement of *E. articulatum* to areas of weak currents and weak tidal disturbance.

One factor which controls the depth at which foraminiferids can live in the sediment is the thickness of the oxidized surface layer. In the Jade area, in common with other tidal flats, a thin oxidized layer of surface sediment overlies the main mass of anaerobic sediments. Richter (1964b) found the oxidized surface layer to be thinner in summer than in winter, and thicker in sandy than muddy sediments at any given season. It was thinner at high-water mark than at low-water mark.

In thin oxidized surface layers, *P. anglicum* and *E. articulatum* flourish, but *E. excavatum* does not because it cannot live at its normal depth if the sediment is anaerobic.

Boltovskoy (1966) reviewed this problem and described observations on cores from Deseado Creek, Argentina. Specimens with protoplasm were found at a maximum depth of 16 cm but they were rare. In the range 12–14 cm depth, 12 species were present; at 4–6 cm, 22 species; and at 0–2 cm, 28 species. Sandy sediments were penetrated more deeply than muddy ones, and Boltovskoy suggested that this was related to the better aeration of the former.

The almost uniform distribution of living individuals in the top 3 cm of surface sediment in Narragansett Bay led Brooks (1967) to conclude that *Ammonia beccarii* (Linné) is infaunal rather than epifaunal.

To test whether some foraminiferids are truly infaunal, Frankel (1972) took cores from pebbly nearshore sands in a Connecticut estuary and immediately preserved them in alcohol to prevent any migration by foraminiferids. They were stained, impregnated and sectioned. Most of the tests were empty. Those that contained protoplasm seemed to be undergoing gamogony and were present mainly between 3 and 9 mm depth. Frankel regarded this as proof of an infaunal habitat.

Furthermore, Sheehan and Banner (1972) found live *Elphidium incertum*

(Williamson) in moist mud from beneath the dried-out mud surface of a tidal marsh.

Perhaps the most unequivocal example is that described from the subsurface of the Kara-Kum desert (Nikoljuk, 1968). Three species were found: '*Miliammina oblonga* (Montagu) var. *arenacea* Chapman', *Spiroloculina turcomanica* Brodskij and *Fischerina* sp. All were very small and thin-shelled, but different growth stages were present and individuals containing protoplasm were stained with boro-carmine. They live in the interstices between the sand grains, in interstitial water with a salinity often greater than 30 per mille and a temperature of 17 to 20 °C. They are thought to be relict from the Pliocene sea which covered the area.

Lee *et al.* (1969) have pointed out that the upward migration of foraminiferids in sediment can be used to retrieve living individuals from sediment samples. The sediment is placed in a bowl, and after some hours the foraminiferids have climbed to the surface and often on to the sides of the bowl. They experimented with twenty-five *Ammonia beccarii* (Linné) and *Protelphidium tisburyensis* (Butcher) which they buried to a depth of 1 to 8 cm beneath sediment. The area below the sediment surface was kept dark. After four days, 80 per cent of the *A. beccarii* but none of the *P. tisburyensis* had reached the surface. By contrast Marszalek *et al.* (1969) observed *Peneroplis* to burrow to a depth of 2 cm in sand in their culture tanks.

These results have great relevance to sampling methods, especially if some species have a preferred depth zone. However, in the author's experience, the difficulties involved in the dilution of living individuals by dead forms in samples greater than 1 cm thick, is more troublesome than any theoretical increase in the sample validity that might arise from thicker samples.

EXPERIMENTS ON TRANSPORT OF LIVING ANIMALS

Richter (1965) recorded that *Elphidium articulatum* (d'Orbigny) (= *E. excavatum* of Richter) in his cultures sometimes extended their pseudopodial net and attached themselves to the under-surface of the water/air interface. In this state of being suspended from the surface, they wandered round the culture dish for some days before sinking once again to the bottom. Under natural conditions the animal would be in contact with the water surface at low tide and, if the rising waters of the flood tide were not turbulent, it would be able to remain attached to the surface and become suspended. In this condition it could easily be transported before dropping to the sea floor. Similar behaviour was reported by Richter in the snail *Hydrobia ulvae*. Neither animal is capable of swimming.

Richter believed that part of the enrichment of the standing crop at high-water mark in the Jade Bay of Germany is due to transport of living *Protelphidium anglicum* Murray (= *Nonion depressulus* of Richter) and *Elphidium articulatum* (d'Orbigny) (= *E. excavatum* of Richter). The other dominant species, *Elphidium excavatum* (Terquem) (= *E. selseyense* of Richter), which lives deeper in the sediment, is not affected by transport processes.

MORPHOLOGICAL VARIATION

This arises in two ways: as a result of reproduction and as deformity. Under the first heading would be included differences due to microspheric and megalospheric generations and to the normal variability of form encountered in healthy populations.

Extreme variability has been recognized in cultures of tectinous foraminiferids (see review by Hedley, 1964) and in the calcareous species *Cibicides lobatulus* (Walker and Jacob). Nyholm (1961) found a great variety of test shapes resembling the genera *Crithrionina* or *Webbina*, *Cibicides*, *Dyocibicides*, *Annulocibicides*, *Cyclocibicides*, *Stichocibicides* and *Rectocibicides*. The forms resembling *Crithrionina* or *Webbina* are a schizont stage from which develop calcareous *Cibicides* types. Less extreme variation has been described in *Rosalina leei* Hedley and Wakefield (1967).

Myers (1943a) recorded size differences between populations of *Elphidium crispum* (Linné) living in the intertidal and subtidal zones of Plymouth Sound. The latter were 60 per cent larger and had 40 per cent more chambers than the former. Further, slow-growing tests were thicker and more circular in outline. Rapid growth led to thinner tests and the development of a more distinct keel.

Under favourable environmental conditions, young individuals will grow to reproductive maturity before a large test size is reached. Schnitker (1967) established cultures of *Triloculina linneiana* d'Orbigny from juveniles of parents originating from Largo Sound, Florida. The parents were large (0.65–1.0 mm long), ribbed forms with eleven chambers. The juveniles in culture reproduced when 0.125 mm long and at the four- to five-chamber stage. They lacked ribs and all were spiroloculine or quinqueloculine; none was triloculine. Schnitker concluded that the favourable environmental conditions in his cultures permitted reproduction to occur before the specific characters had been developed.

Different generations of the same species have sometimes been given different specific and generic names, and this caused inaccuracies in ecological studies. It is doubtful whether these are of great significance from the geological point of view.

Variation due to deformity can be seen in *Cibicides lobatulus* (Walker and Jacob) and other species which attach themselves to their substrate; the attached side of the test commonly takes on the shape of the substrate. An unusual case has been described for *Gypsina* by Nyholm (1962). Some of these had no protoplasm in the chambers of the attached side, although the inner chambers contained dense protoplasm. Nyholm concluded that attachment was not the work of pseudopodia. A further observation was that large discoid *Gypsina* had several groups of juvenile chambers. The interpretation seems to be that a composite adult has grown from a number of closely crowded juveniles which fused on contact. If true, this raised interesting physiological problems. Other complex shell shapes were attributed to the development of new *Gypsina* individuals on the empty shells of former adults killed by sporozoa. Further, *Gypsina* was shown to be a resting stage of *Cibicides*.

Nubecularia lucifuga Defrance from blades of *Posidonia* were cultured by Arnold (1967). The orientation of the prolocular apparatus during settlement controls the subsequent shape of the adult test. There is a great deal of morphological variation, and Arnold considers that half of the modern 'species' belong in *N. lucifuga*. There is great similarity with Jurassic representatives.

Deformity of the test can result when foraminiferids suffer periods of food shortage. Murray (1963) produced foraminiferids with notches in the periphery caused by the addition of small chambers during times of food shortage. Civrieux (1968) attributed this type of deformity to unfavourable salinity but did not support this with experimental evidence.

Heron-Allen and Earland (1910) noted the occurrence of a half-inch (1 cm) thick deposit of foraminiferid shells on the beach at Bognor, Sussex, in March 1910. The majority were of *Massilina secans* (d'Orbigny). To investigate problems of variation in test shape, they collected scrapings from algae on the Mixon Reef and Beacon Rocks offshore. The living foraminiferids were stored in tanks and were observed to multiply in number. In September, the contents of the tanks were dried and examined. Most of the *Massilina secans* specimens were deformed, and included the three varieties described as var. *denticulata*, var. *tenuistriata* and var. *obliquistriata*. Some specimens showed the features of all three varieties. The authors concluded that the varietal forms resulted from living 'under starved and unnatural conditions' (Heron-Allen and Earland, 1910, p. 695). They also noted that damaged shells were repaired by the secretion of new shell by the protoplasm extruded at the break.

Boltovskoy (1963b) described how specimens of *Quinqueloculina seminulum* (Linné) from Patagonia had tried to grow shell material over the egg cases of turbellarians which had used the foraminiferids as a substrate for attachment. Sometimes the egg cases break off, leaving irregular depressed scars on the shell surface. Such egg cases are not uncommonly attached to shallow-water foraminiferids along the south coast of Britain. It is of particular importance that presumably the epizoans attach only to bare shell and therefore the living foraminiferid must withdraw its protoplasm into the shell. Whereas it has long been known that the Miliolina do not have a protoplasmic cover over their shell, it has been less commonly realized that this is true of the Rotaliina too.

Most deformities resulting from starvation or other environmental conditions are easily recognizable and assist the understanding of ecology.

TEMPERATURE

Temperature has long been known to be an important environmental control in limiting the geographic distribution of species and in affecting the rate of the physiological activity.

Orton (1920) defined five critical temperatures:

1. maximum temperature for survival,
2. maximum temperature for successful reproduction,

3. optimum temperature,
4. minimum temperature for successful reproduction,
5. minimum temperature for survival.

Bradshaw (1961) has elaborated this scheme and Civrieux (1968) has suggested a diagrammatic method of representation. Reproduction occurs only within a narrow temperature range which is different for each species. Growth proceeds in a broader temperature range but stops close to the limits which cause death.

Bradshaw (1955) cultured *Rotaliella heterocaryotica* Grell and *Rotaliella* sp. at two temperatures, 14.5 °C and 22.2–24.5 °C. Each species showed a slower rate of population growth at the lower temperature. However, clearly both temperatures were within the limits of temperature for successful reproduction of these two species.

In a second series of experiments using *Ammonia beccarii* var. *tepida* (Cushman), Bradshaw (1957) established cultures of 6–10 individuals, kept them at several different temperatures, and in water of salinity 33.5 per mille. At 10° C no chambers were added. At 15 °C, one new chamber was added every 6 days. At 20 °C, it took 4 days to form a new chamber. In the temperature range 24–30 °C, a new chamber was added every 2 days. Growth ceased at 35 °C, and the specimens died.

The results for *Ammonia beccarii tepida* showed that 10 °C was too low for reproduction but above the minimum temperature for survival. When the culture was warmed to a temperature of 24–27 °C after 27 days without growth, the specimens grew and reproduced after a further 25 days. The 15 °C culture was also below the minimum temperature for reproduction. Reproduction took place at 24–27 °C on average every 19 days, while at 30 °C it averaged 15 days. Presumably this latter temperature is close to the optimum. Death occurred at 35 °C indicating that the maximum temperature for survival had been exceeded.

A further result from Bradshaw's experiments was that specimens reproducing at lower temperatures had more chambers and a greater mean size than those that reproduced at higher temperatures. Bradshaw concluded that, even though a foraminiferid may have reached maturity, it would only reproduce if the environmental conditions were favourable. This is particularly important in the case of foraminiferids, as reproduction effectively means the death of the parent.

A continuation of the experiments (Bradshaw, 1961) was concerned with lethal and reproductive temperatures. Several species were used to determine the lethal limits. *Ammonia beccarii tepida* from three localities all died at 45–46 °C, *Bolivina compacta* Sidebottom at 41 °C, *Massilina* sp. at 40 °C, *Spirillina vivipara* Ehrenberg at 39 °C, *Rosalina columbiensis* (Cushman) at 39 °C and *Bolivina vaughani* Natland at 38 °C. Whereas the previous history of exposure to temperature and salinity changes prior to the experiments did not seem to affect the results much, long periods of exposure caused death at lower temperatures than the maximum values recorded above.

Ammonia beccarii tepida was again used to determine the effects of temperature on reproduction. The minimum temperature for successful reproduction was 20 °C and the maximum was 32 °C, but the optimum was 25–30 °C. The rate of reproduction averaged 88 days at 20 °C to 33 days at 30 °C. At 30–32 °C, the reproductive period became longer. The growth rate showed a similar pattern with a minimum of 4 μm per 10 days at 15 °C to a maximum of 84 μm per 10 days at 30 °C, followed by a fall above that temperature. Again the largest tests formed under the coolest conditions.

SALINITY

Changes in salinity affect the osmosis of foraminiferids. It is generally assumed that microscopic marine animals are isotonic with sea water, and they are therefore described as poikilosmotic. Fresh-water animals are hypotonic to their environment and are described as homoiosmotic. Brackish-water animals can fall into either category. Homoiosmotic animals maintain their osmotic pressure by osmoregulation, whereas poikilosmotic animals have no need to do so.

It is a matter of observation from the distribution of living foraminiferids, that some species are restricted to truly marine waters (stenohaline) while others tolerate hyposaline or brackish-water conditions for some or all the time (euryhaline).

The earliest systematic experiments involving the effects of salinity on foraminiferids were carried out by Bradshaw (1955). Using *Rotaliella heterocaryotica* Grell, cultures were prepared at salinities ranging from 16.8 to 83.7 per mille. Growth of the populations was measured over a period of 67 days. Increase in population took place in cultures having salinities of 26.8 per mille to 33.5 per mille. However, at salinities less than 23.5 per mille and more than 37.0 per mille no growth took place.

Ammonia beccarii tepida (Cushman) was used by Bradshaw (1957) in a second set of experiments. Cultures of 6–10 individuals were kept at a temperature of 24–27 °C. Waters of different salinity were used and the growth of individuals was measured over a period of up to 60 days. At salinities of 20–40 per mille it took 2 days to add a new chamber. At salinities of 13 per mille and 50 per mille, a new chamber was added every 3 days. A culture kept at 67 per mille salinity showed no growth but survived. Cultures at salinities of 7 per mille showed slight growth but those at 2 per mille did not grow. However, in each case the specimens stayed alive. Similar results were obtained by Bradshaw (1961).

Reproduction took place every 23 days at salinities of 20–40 per mille, and every 48 days at 13 per mille. No reproduction occurred when the salinity exceeded 50 per mille. Specimens reproducing near the limits of favourable salinities commonly had more chambers and were slightly larger than those reproducing under optimum conditions. This parallels the observations made on the effects of temperature on reproduction.

Lethal salinities were determined for several species by Bradshaw (1961): *Ammonia beccarii tepida* less than 2 per mille, *Bolivina vaughani* 3 per mille, *Massilina* sp. 7 per mille, *Spirillina vivipara* 7 per mille. However, it was noted that *A. beccarii tepida* could withstand exposure to distilled water for up to 32 minutes.

Experiments on *Elphidium crispum* (Linné) carried out by Murray (1963) showed that the rate of feeding was related to the salinity. At a salinity of 20 per mille or less, no feeding took place. In a second experiment the calcium content of the subsaline waters was adjusted to that of normal sea water, but the rate of feeding remained unaltered and feeding had ceased by 20 per mille salinity. The results were the same in a third experiment in which calcium was excluded from the water. Thus, the presence or absence of calcium had no effect on the rate of feeding; this was controlled only by the salinity of the medium.

The endurance of *E. crispum* to lowered salinities was tested as follows; 8 cultures each of 10 healthy individuals were set up in two series. The first series of 4 cultures was kept at 8 °C, the winter mean temperature for the mouth of the English Channel, and the second at 16 °C, the summer mean temperature. In each series, two cultures were of 20 per mille and two of 15 per mille salinity, selected because in previous experiments the species had not fed at these salinities. As might be expected for cultures of the same salinity (20 per mille) those kept at 16 °C fed faster than those at 8 °C. However, the individuals kept at 8 °C remained healthy and fed when returned to normal sea water after 38 days, while fewer of those kept at 16 °C did so. No feeding took place in any of the cultures kept at 8 °C, but after the return to normal salinity on the 15th–38th day for the two 8 °C cultures, feeding was resumed by a few individuals. Specimens kept in 15 per mille salinity and at a temperature of 16 °C were apparently dead. It was concluded that for a marine species, survival in water of lowered salinity was improved if accompanied by low temperatures. This may be due to the relationship between osmotic pressure and temperature. The osmotic pressure of salinity 20 per mille is 14.19 atmospheres at 25 °C (Robinson, 1954). Using the conversion factor

$$1 + \frac{t - 25}{298}$$

where t (°C) is another temperature, the result is 13.76 atmospheres at 16 °C and 13.38 atmospheres at 8 °C. The difference is small and it may be that temperature is more important than osmotic pressure because it slows down the metabolic processes of the foraminiferids.

pH

The only experimental work showed that at pH 2.0 *Ammonia beccarii tepida* (Cushman) survived 25–75 minutes, compared with 2–6 minutes for *Spirillina vivipara* Ehrenberg (Bradshaw, 1961). However, at sublethal low pH values, the calcareous tests were dissolved without harming the animals and new tests were secreted on return to more favourable conditions.

Myers (1943a) noted that *Elphidium crispum* (Linné) fell off the sides of the aquarium when the pH fell below 7.3. This indicates some loss of control of the pseudopodia even if only as a shock reaction. Boltovskoy (1963c) recorded the death of *Elphidium macellum* (Fichtel and Moll) and *Quinqueloculina seminulum* (Linné) at pH 7.0.

HYDROSTATIC PRESSURE

The only experiments designed to test the effects of hydrostatic pressure were those carried out by Bradshaw (1961). He used *Ammonia beccarii tepida* (Cushman), a lagoonal species, and tested it to pressures of 1000 atmospheres. High pressures could be withstood for short periods only. This indicates that shallow-water foraminiferids could not exist in the abyssal depths of the ocean and that the foraminiferids there must be specially adapted to withstand the pressure.

PREDATION AND DEATH

Predation on foraminiferids has been discussed recently in a review by Sliter (1971), and in a consideration of trophic structure of marine communities by Lipps and Valentine (1970). Predators are of two types, unselective and selective.

Unselective predation on benthic foraminiferids results from the activities of deposit feeders. These ingest sediment and, since foraminiferids are too slow to escape, they must also be eaten. Animals from which foraminiferids have been obtained include worms, flatworms, sea urchins, holothurians, gastropods and bivalves (Myers, 1943a; Mateu, 1968; Morton, 1959a).

Selective predation has been observed in the following animals: *Gonoplax* (Mare, 1942), shrimps (Myers, 1943a) and decapods (Orton, 1927; Sokolova, quoted by Lipps and Valentine, 1970). Myers noted that shrimps feed on foraminiferids mainly during periods of scarcity of other food. Predatory molluscs include various opisthobranch gastropods and scaphopods (Morton, 1959b; Graham, 1955). In most cases the foraminiferid tests pass unharmed through the gut of the predator, but the scaphopods and some gastropods grind them up.

Nematodes have been reported within foraminiferid tests by Myers (1943a) and Sliter (1971). Myers considered them to be parasites. Sliter regards them as predators. He carried out experiments to determine whether they bore holes in the foraminiferid tests. Although he did not observe any attacks by the nematodes, the constant association of holes in the test of *Rosalina* and the presence of nematodes in the chambers is fairly conclusive. The holes are 3.7 to 14.3 μm in external diameter, and round to oval in shape. They occur mainly on the dorsal side in the ultimate or penultimate chamber. Similar borings are found in the apertural end of *Bolivina doniezi*. The abundance of bored tests varies from 3 per cent in shallow estuarine waters to 16 per cent at a depth of 2000 m in the Gulf of Mexico, but there is no clear correlation between depth and predation.

Sliter considers that predator abundance is generally associated with prey density. Although it is claimed that: 'predation can directly affect estimates of foraminiferal production' this is not necessarily true. It is well known that many progeny must die before they are old enough to reproduce, otherwise foraminiferid standing crop would increase at an alarming rate. Predation is just one method of being killed, and since the tests are only bored and not destroyed, they are still added to the sediment.

Nyholm (1961) has reported that sponges grow over *Cibicides lobatulus* (Walker and Jacob) living on ascidians in the Gullmar Fjord. The cytoplasm is then destroyed. This is probably an unusual method of death and may not be related to feeding by the sponge.

Reproduction, whether sexual or asexual, results in the parent test being left free of protoplasm. Thus the animal has died in the sense that its empty test is added to the sediment. In some cases it is possible to recognize this mode of death due to special morphological structures associated with reproduction, e.g. float chamber in *Tretomphalus*.

There is no published evidence for disease in foraminiferids, although death from this source cannot be ruled out. Many forms in marginal marine areas are thought to die through rapid changes in environmental conditions beyond the limit tolerated by individual species. Sometimes storms uproot weeds on which foraminiferids live and transport them into an unfavourable environment, e.g. a beach, and this causes death.

All the methods of dying discussed cause the foraminiferid tests to be added to the sediment. The only exception is where certain molluscs grind up the tests and so destroy them.

20. *Ecological Controls*

GENERAL STATEMENT

Ecology is the study of the interrelationships between the environment and organisms. Abiotic aspects include temperature, salinity, oxygen etc. Biotic aspects include the relationships between animals and plants.

Most studies of foraminiferids have been concerned with species and population ecology. The place of foraminiferids in communities is discussed in Chapter 17. Their importance in the ecosystem has scarcely been studied in detail.

This chapter is concerned with environmental parameters that can be regarded as limiting factors. For each of these parameters each species has a certain range of tolerance:

minimum for survival		range of tolerance for survival
minimum for reproduction	range of tolerance for reproduction	
optimum		
maximum for reproduction		
maximum for survival		

The recognition of different tolerance levels for survival and reproduction is important. A species may be able to *survive* in an environment in which it is *unable to reproduce*. In such cases the continued existence of the species is precarious, for it is dependent on an environmental change to conditions favourable for reproduction. Nevertheless, many temperate species are dependent on such changes, and reproduction takes place only at those times of the year when conditions are favourable.

The range of tolerance concept applies to all limiting environmental factors. A species will survive only if none of these factors exceeds the range of tolerance for survival. Even if only one factor exceeds the limit for survival, the species will die off.

Most studies of benthic foraminiferids have insufficient ecological data to determine the true cause of distribution patterns. It is apparent from the chapters on distribution that populations are affected not only by the size of environmental change, but also by the speed. In marginal marine environments, changes are large and fast. In shelf seas, changes take place slowly so that there is more chance that the organisms will be in equilibrium with their environment.

The tolerance limits of individual species can be established either by detailed long-term field studies or by experiment with laboratory cultures. Only the latter has so far been attempted, and the results are discussed in Chapter 19. It follows from these remarks that (*a*) at any one locality, different environmental parameters will limit feeding, reproduction or survival at different seasons of the year, and (*b*) in field studies carried out over a few days, it will be possible to recognize only the factors close to the tolerance limit at that particular time. This explains why different studies of similar environments reach different conclusions about the rôle of individual environmental parameters. For this reason no detailed comparisons of different study areas is given in the following brief notes (for details see Boltovskoy, 1963c).

ABIOTIC

Temperature

Foraminiferids are poikilothermic. Some are eurythermal, i.e. have a wide tolerance, while others are stenothermal, i.e. have a narrow tolerance. In individual areas of sea floor, both the annual and diurnal ranges are important for individual species distributions. Reproduction occurs most frequently at the optimum temperature (all other factors being favourable) and more slowly close to the tolerance limits. On a worldwide basis, the major shallow-water faunal provinces are to a large extent controlled by the distribution of temperature differences. The correlation of decrease in temperature with increase in depth makes it difficult to determine the true cause of 'depth' zonation.

Salinity

The majority of benthic foraminiferids are believed to be stenohaline, i.e. they will tolerate only small changes of salinity. Some species are euryhaline and these forms are found in marginal marine environments. Some euryhaline species live in hyposaline and hypersaline waters. The majority of hyposaline species are confined to that environment and rarely occur in normal marine conditions. By contrast, all hypersaline species also live in normal marine conditions.

Changes in salinity affect the density of the water and its osmotic effect on the foraminiferids. The clearest control by salinity is its effect on miliolaceans. At salinities less than 32 per mille, these are unable to maintain their pseudopodial reticulum. Consequently they do not live in hyposaline waters.

The importance of salinity as a limiting factor is emphasized throughout the chapters on distribution in the different environments.

Availability of calcium carbonate

The solubility of calcium carbonate is controlled by the salinity, temperature and carbon dioxide content of the sea water. In general, calcium carbonate is most readily available in tropical marine or hypersaline waters, and least easily available in cool hyposaline waters. Species with agglutinated tests and non-calcareous cements are dominant in regions of low calcium carbonate availability, e.g. some tidal marsh-pools. Once the level of calcium carbonate availability is reached, it is probably not a limiting factor.

Depth

In many foraminiferid studies, depth is the only physical parameter measured. However, as Funnell (1967) has pointed out, depth in itself is probably not a limiting factor. Many parameters are related to depth: hydrostatic pressure, density, light penetration, temperature, pH, oxygen content and carbon dioxide content. Although there may be a correlation between depth and the distribution of foraminiferids in restricted geographic areas, on a worldwide scale individual species do not appear to be controlled directly by depth alone.

pH

This has not been recorded as a limiting factor in the marine environment. In marshes, low pH causes postmortem solution of tests although it appears not to affect the living animals.

Oxygen

The lower limit tolerated by most organisms is $1-2$ cm^3/l (Emery and Stevenson, 1957). Foraminiferids do not live in anoxygenic conditions such as in the deep basins of the Baltic (Lutze, 1965). Otherwise, oxygen is not recorded as a limiting factor.

Illumination

There is no clear evidence that availability of light directly controls the distribution of foraminiferids, although those species with symbiotic algae would gain from good illumination. A number of species live in association with algae and sea-grass which are themselves controlled partly by submarine illumination. Indirectly, illumination also affects the growth of foraminiferids through algal (photic zone) or non-algal (aphotic zone) food supply.

BIOTIC

Food

Most of our knowledge of the food requirements of foraminiferids comes from experimental studies (see Chapter 19). Field studies of dispersal in the intertidal zone show correlation of maximum standing crop size with greatest availability of food (Lee *et al.* 1969). Few authors have attempted to measure the availability of food, yet it must be of fundamental importance in determining the standing crop and biomass.

21. Relationship Between Living and Dead Assemblages

Usually, the living and dead assemblages in the same sample differ in the relative abundance of individual species, and sometimes there are additional species in the dead assemblage. These differences arise partly from production and partly from postmortem changes.

PRODUCTION

In most assemblages, some species reproduce more frequently than the rest. Consequently they make a larger contribution to the annual production and therefore to the dead assemblage. Table 25 gives data for a theoretical ex-

Table 25 Data for standing crop and annual production in a theoretical assemblage

Species		Standing crop Number	%	Annual production Number	%
Textulariina	A	50	22.8	50	9.8
	B	50	22.8	100	19.5
Miliolina	C	50	22.8	150	29.5
Rotaliina	D	40	18.0	120	23.5
	E	30	13.6	90	17.7
Totals		220	100	510	100
% Textulariina			45.6		29.3
% Miliolina			22.8		29.5
% Rotaliina			31.6		41.2
α index		<1		<1	

See page 204 for further data

ample. The percentage abundance of each species is different in the standing crop and the resultant death assemblage. Nevertheless, the similarity index between the two is 83.7 per cent. On a triangular plot, they have different positions (Figure 98 (*a*)). Examples from the literature show enrichment in one component: Rotaliina (Figure 98 (*b*)), Miliolina (except for G) (Figure 98 (*c*)), and Textulariina (Figure 98 (*d*)). In each restricted sampling area, the same relative difference exists between the living and dead assemblages, and this is what one would expect if the cause is due to production differences.

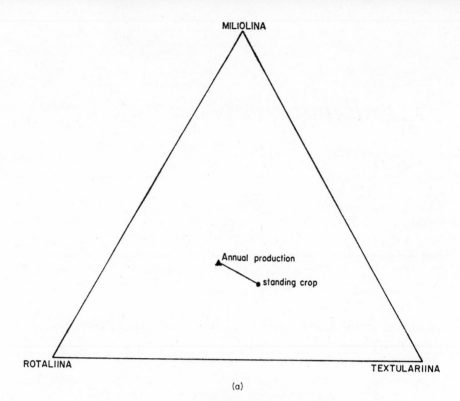

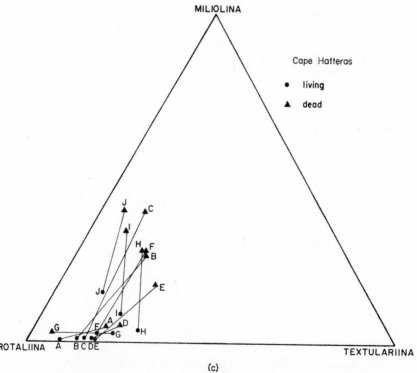

Figure 98 (Part 1) (*a*) triangular plot of standing crop
and annual production, based on Table 25.
(*b*) live–dead Vineyard Sound.
(*c*) Cape Hatteras,
(*d*) shelf off Long Island (data from Murray, 1969).

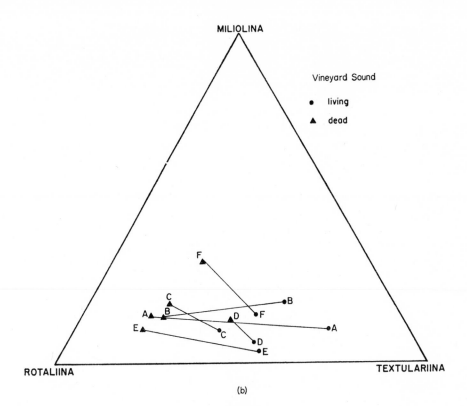

(b)

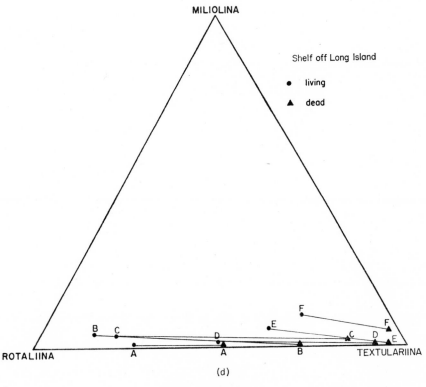

(d)

Figure 98 (Part 2)

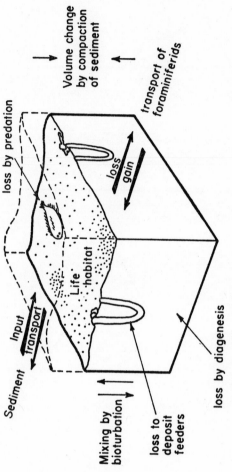

Figure 99 Postmortem processes operating on foraminiferids.

POSTMORTEM CHANGES

Some of the postmortem processes which operate on benthic foraminiferids are shown diagrammatically in Figure 99. These include transport, bioturbation and diagenesis.

Transport

Foraminiferids can be transported by water (as suspended load or bed load), floating plants, ice, turbidity currents and wind.

Suspended load Empty tests of twenty-seven species of benthic foraminiferids were reported from plankton tows in the English Channel by Murray (1965d). The samples were taken from three stations at the surface and at a depth of 10 m. Prior to sampling there had been a long period of storms. All the foraminiferids were size sorted, and most were 0.15 to 0.20 mm in diameter or length. Only thin-walled types were present. Lidz (1966) found *Bolivina vaughani* Natland in plankton tows from off California. He assumed that it was planktonic during part of its life cycle. Loose (1970) recorded twelve species, living and dead, from off California. The size range was 0.25 to 0.50 mm diameter. A cluster analysis comparison of this assemblage with that from the adjacent intertidal zone showed them to be indistinguishable. He suggested that the foraminiferids are washed out of tide pools and transported within the white foam of breaking waves. Foraminiferids transported in this way have been recorded several hundred kilometres offshore. This is one method of introducing shallow-water foraminiferids into deep-water sediments. It may account for the disparity observed in depth ranges of living and dead representatives of the same species.

When tidal flats are exposed to the sun, foraminiferid tests dry out and become air-filled. The incoming tide picks up such light shells and transports them shorewards. This type of transport is common in Abu Dhabi lagoon where the hypersaline, and therefore denser, water makes it easier for the tests to float. However, the process also operates in estuaries and was reported from the Dee by Siddall (1878).

Bed load Transport as bed load must be very common, especially in shallow-water areas subject to currents. The factors which affect transport are size, specific gravity of the shell in sea water, and shape. Because foraminiferid tests are hollow, their excess weight in sea water is much less than that of a quartz grain of comparable size.

Jell *et al.* (1965) determined the specific gravity of dry reefal foraminiferids. Complete *Calcarina* has a mean value of 1.85, whereas eroded specimens have a mean of 2.2; for complete *Baculogypsina* the mean is 1.9 and for *Marginopora* it is 2.3. All other bioclasts had values in the range 2.4 to 3.0, so foraminiferids are potentially more liable to transport. Plates such as *Marginopora* settle significantly slower than other shapes (Maiklem, 1968). Spheres like *Calcarina* and *Baculogypsina* also settle slowly because of their low density. *Alveolinella*, a rod, settles fastest of all (Figure 100). All these particles settle much slower than the equivalent-sized solid calcite particle.

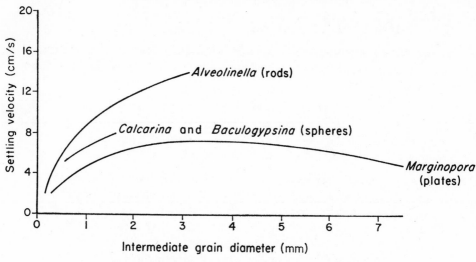

Figure 100 Setting rates of foraminiferids (data from Maiklem, 1968).

Haake (1962) and Grabert (1971) have made experimental studies of the settling velocity of small intertidal foraminiferids. Their results are reproduced in the following table.

Species	Diameter (mm)	Equivalent diameter (mm)	Nominal diameter (mm)	Length/ breadth ratio	Settling velocity (m/s)
Elphidium excavatum					
(plump form)[1]	0.35	0.16	0.31	2.15	0.0160
(flat form)[1]	0.35	0.10	0.299	2.66	0.00915
Protelphidium[1]					
anglicum	0.35	0.125	0.314	2.42	0.0119
Trochammina[2]	0.43	0.135	0.31	2.6	0.011
Eggerella scabra[2]	0.385	0.16	0.243	2.17	0.0179

1. Haake (1962);
2. Grabert (1971); *E. excavatum* = *E. selseyense* of Haake; *P. anglicum* = *Nonion depressulum* of Haake; settling velocity in distilled water at 20 °C.

The settling velocity and the equivalent diameter of a quartz grain are dependent on the size, shape and weight of the foraminiferid. Grabert found that quartz grains and *Trochammina* with the same hydraulic equivalent diameter have a size ratio of roughly 1:3. However, in Bottsand Lagune, *Trochammina* and sand grains of equivalent size do not occur together; the foraminiferids are found with finer sand. In the Langeoog tidal flats Haake (1962) considered that, as the distribution of living and dead foraminifers is essentially the same, transport has not led to size sorting.

Experiments carried out by Nichols and Norton (1969) show that *Elphi-*

dium clavatum Cushman settles twice as fast as *Ammonia beccarii* (Linné) and two to three times as fast as *Ammobaculites crassus* Warren.

Turbidity currents and slides Downslope movement of sediments as turbidity currents or slides is known from oceanic and shelf sea basins. Field observations by Schafer and Prakash (1968) using a sediment trap in Bedford Basin, Nova Scotia, showed that the turbid bottom water contained ten species of foraminiferids, seven of which had live representatives. The size range was 0.14 to 0.66 mm.

Shallow-water intercalcations in deep-sea cores are easily recognized on lithological and faunal evidence. Sometimes there are mixed assemblages of deep- and shallow-water species. Phleger (1960a) suggests this may be due to colonization by deep-water species of the top of a newly introduced layer of shallow-water origin.

Vegetation Foraminiferids living on sea-grass or seaweeds in shallow water are liable to be transported when periodic storms uproot the vegetation. Often, transport is shorewards.

Ice In the Jade area of Germany, ice develops in the intertidal zone. The surface sediment layer and its included foraminiferids are frozen into the lower surface of the ice, which is transported shorewards by the rising tide (Richter, 1965).

Wind Onshore winds remove material from the intertidal zone and transport it on to the adjacent land. In Dogs Bay, Connemara, Ireland, there are dunes composed of bioclastic material including abundant *Cibicides lobatulus* (Walker and Jacob). This species lives on seaweeds attached to rocks in the adjacent nearshore zone. Along the Trucial Coast of the Persian Gulf subaerial dunes up to 12 m high form along shores where there are sublittoral oolith deltas. The foraminiferid fauna of the dunes is essentially the same as in the subaqueous oolitic sediment (Murray, 1970a).

Bioturbation

Most sedimentary substrates support an infauna and an epifauna. The infauna dig burrows and, by their feeding activities, ingest sediment from one point and defaecate it elsewhere. Primary depositional layering becomes destroyed and foraminiferids from different levels become mixed. Providing the dead foraminiferids are the same as those living in the surface, bioturbation causes little alteration of the assemblages. Brooks (1967) has suggested that *Ammonia beccarii* (Linné) is infaunal in the top 4 cm of sediment, and this may be true of other benthic foraminiferids.

Relict assemblages

The widespread occurrence of relict sediments on the continental shelves of the world has recently been reviewed by Emery (1968). As much as 70 per cent of the shelf area may have only the thinnest veneer of recent deposits over relict sediment. This is important in the study of benthic foraminiferids.

Because of the present non-deposition, the top of the relict sediment represents a condensed horizon rich in foraminiferids which are partly modern
and partly relict. The diversity is often abnormally high. Under normal
conditions of sediment accumulation, this mixing of assemblages does not
occur.

Relict sediments are not found in lagoons and estuaries because they are
shallow and the site of fairly rapid deposition.

Derived fossils

It is not uncommon for fossil foraminiferids to be eroded out of soft sediments such as clays, marls and chalks, and transported into a modern
depositional environment. Sometimes the foraminiferids are badly abraded
but small planktonic species may remain in a near perfect condition. Older
foraminiferids are often different in colour and lack the fresh appearance of
recent forms. Where subrecent deposits are eroded, it may not be so easy to
recognize reworked specimens. Some of the many references to this topic
are Resig (1958), Zalesny (1959), Uchio (1960) and Bandy (1964).

Diagenesis

In soft sediments, the most serious diagenetic change is removal of calcareous
shells by solution. In extreme cases all foraminiferids are lost. Partial
solution is indicated by white, opaque, and etched tests of hyaline foraminiferids (Murray, 1967a; Murray and Wright, 1970). Agglutinated forms can
also be destroyed by solution of calcareous and organic cements.

22. Summary of Foraminiferid Distributions and their Application to Palaeoecology

SUMMARY OF DISTRIBUTION DATA

Diversity

The summary of the ranges of diversity (Figure 101) clearly shows that it is possible to differentiate environments. In a general sense, $\alpha = 5$ is a boundary separating normal marine environments ($\alpha > 5$) from abnormal environments ($\alpha < 5$). Hyposaline and hypersaline marshes, lagoons and hyposaline shelf seas all have low diversity. Normal shelf seas and normal marine lagoons have diversity values of $\alpha > 5$.

Triangular plot

The fields for different environments show some overlap, but on the whole there is a very clear pattern (Figure 102). The field for hypersaline marshes occupies the entire triangle. *All* hyposaline environments lie close to the Rotaliina–Textulariina side: hyposaline marshes lack Miliolina, hyposaline lagoons and estuaries have a small Miliolina component, especially in their seaward parts. Shelf seas have up to 20 per cent Miliolina, although many lack the group altogether. Normal marine and hypersaline lagoon assemblages are primarily mixtures of Miliolina and Rotaliina.

Genera

In Appendix 2, ecological data are listed for eighty-three genera: salinity, substrate, temperature of the bottom water, and depth. Some genera are good environmental indicators; however, different species of some genera live in totally different environments.

APPLICATION TO PALAEOECOLOGY

This book has been written primarily to assist the palaeoecological interpretation of fossil foraminiferids. The more that is known of the distribution and ecology of modern species, the easier it will be to interpret fossil assemblages. The value of environmental reconstructions based on foraminiferids has been apparent to the petroleum industry for many years. It has been a major factor promoting the development of the subject.

Before attempting to interpret a fossil assemblage, it is wise to note the

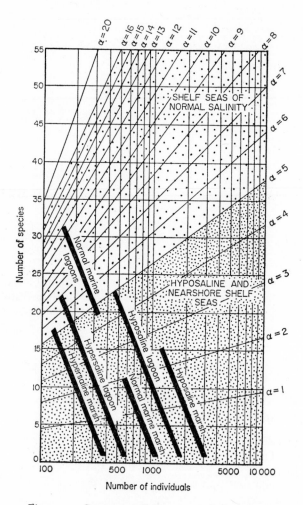

Figure 101 Summary of the range of diversity in different environments.

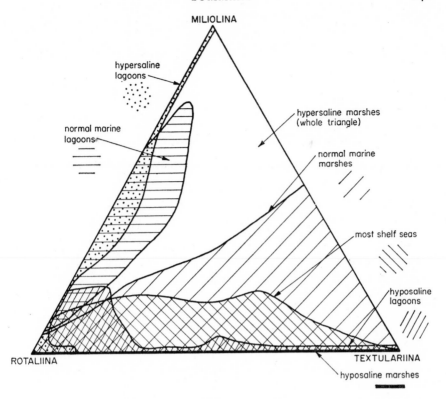

Figure 102 Summary triangular plot.

geological setting of the sample. In particular, coarse-grained sediments are likely to show evidence of reworking and transport of foraminiferids. It is important to look for signs of abrasion, size-sorting and mixing of assemblages. If any of these are well developed, an ecological interpretation based on the sample is likely to be partly in error.

If the sample shows little sign of postmortem transport effects, the following procedure can be adopted:

1. Make an assemblage count and determine the α index and the Rotaliina–Miliolina–Textulariina ratio.
2. Examine the α summary diagram (Figure 101) to assess the range of environmental possibilities.
3. Examine the triangular plot summary diagram (Figure 102) to see which environment is indicated.
4. Compare the genera with the information in Appendix 2.
5. When the likely environment has been established, refer to the relevant chapter for details of the modern foraminiferids.

Example

Data α = 4, Rotaliina 80 per cent, Miliolina 5 per cent, Textulariina 15 per cent, few abraded specimens, *Elphidium, Ammonia, Miliammina* and *Ammotium* are the dominant genera.

Inference α index suggests (*a*) hyposaline and nearshore shelf-seas, (*b*) hypersaline lagoon, (*c*) hyposaline lagoon.

Triangular plot suggests (*a*) shelf sea, (*b*) hyposaline lagoon, (*c*) hyposaline marsh.

Genera *Elphidium* (unkeeled) salinity 0–70 per mille, sediment and vegetation, 1–30 °C, 0–50 m, hyposaline to hypersaline marshes and lagoons, nearshore.

Ammonia hyposaline to hypersaline, sediment, 15–30 °C, 0–50 m, hyposaline and hypersaline lagoons, inner shelf.

Miliammina salinity 0–50 per mille, sediment, 0–30 °C, 0–10 m, hyposaline lagoons, hyposaline to hypersaline tidal marshes.

Ammotium hyposaline–hypersaline, sediment, 0–30 °C, 0–10 m, tidal marshes and hyposaline lagoons, estuaries, enclosed shelf seas.

Interpretation It is clearly abnormal marine. The low abundance of Miliolina favours hyposaline water. The α index is too high for a hyposaline marsh. It must therefore be a hyposaline lagoon or nearshore shelf-sea. *Miliammina* and *Ammotium* are not found in nearshore shelf-seas, so it

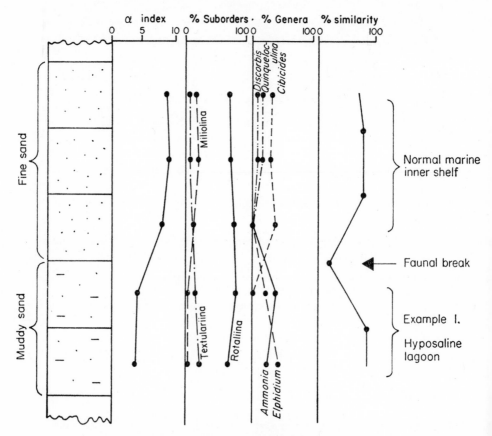

Figure 103 Theoretical borehole record for palaeoecological interpretation.

must be a hyposaline lagoon. The depth is 0–10 m, sediment substrate probably with no weed cover, temperature in the range 15–30 °C in the summer.

Commonly palaeoecological studies are carried out on borehole data. Using the methods described in this book, it is possible to plot a borehole log of α index, percentage occurrence of the suborders, percentage abundance of the dominant genus, and similarity index. The latter is especially valuable in recognizing faunal changes (Figure 103).

The advantages of using several different aspects of the assemblages to deduce the environments they represent, is that the results are less subjective than those obtained by conventional methods. Their application to continuous stratigraphic sections in outcrop or from boreholes enables an assessment to be made of the cause, and therefore the significance, of a faunal change. This is of fundamental importance where benthic foraminiferids are being used for stratigraphic correlation.

Appendix 1. Methods

SAMPLING ERRORS

The importance of collecting and handling samples with the utmost care cannot be too heavily stressed. Ultimately it is the quality of the sample which controls the final result. There is little point in using sophisticated statistical techniques and computer analysis of results if the samples on which these are based were badly collected.

Choice of sampling equipment is important. The desired sample is a known area of sea floor sampled to a shallow depth, e.g. 1 cm. Errors arise due to the pressure wave on impact of some samplers, e.g. grabs, corers, and to washing due to incomplete closure of the sampler during the passage from the sea floor to the deck of the ship, e.g. grabs, dredges. In general, it is the smaller species which will be lost. As an example of the percentage error, if it is supposed that a corer which takes a sample of 10 cm² (diameter 3.6 cm) loses a 1 mm strip all round the edge, due to the impact pressure wave, the area of the sample will be reduced by 10 per cent to 9 cm² (diameter 3.4 cm).

The second problem is whether or not the sample is representative. This depends on two main factors, the size of the individuals and the nature of their distribution on the sea floor. Dennison and Hay (1967) have discussed this problem and applied binomial sampling theory to determine the required sample size. They conclude that a sampler collecting 0.1 m² area of the sea floor (Petersen grab) is adequate for organisms up to 2 mm in diameter. On this basis, samples for foraminiferids should be much larger than the 10 cm² used by many authors.

SAMPLING METHODS

Recent reviews provide fairly detailed information on most of the samplers used for collecting benthic organisms (Holme, 1964; Hopkins, 1964). The ones commonly used in the study of foraminiferids are briefly described below.

Two types of sampler take a sample of known area. The small free-fall corer described by Phleger (1951) has a short, plastic-lined tube with cross sectional area of 10 cm². The top 1 cm of core is normally taken as the sample (volume 10 cm³). A sampler which slices an area of more than 100 cm² from the sea floor has been used by Murray (1969, 1970c). A Van Veen grab can sometimes be used in very shallow water to obtain samples of known area (Murray, 1970b). Van Veen, Dietz-Lafonde, orange peel and Shipek grabs all close on impact with the sea floor. Closure is imperfect, and washing of the material takes place in transit to the deck of the ship. Often, the sample does not represent the whole area of the jaws of the grab and it is therefore of unknown area. Dredges are sometimes used in shallow water, but care must be taken to avoid washing the sample.

Preservation

As soon as a sample has been collected, either 60 per cent ethyl alcohol or 5–10 per cent buffered formalin should be added. Both preservatives cause shrinkage and hardening of the protoplasm. Boltovskoy (1963a) has discussed the effects of formalin on protoplasm and the subsequent uptake of rose Bengal stain.

LABORATORY ERRORS

Several methods of staining living foraminiferids have been tried, but the most successful is that described by Walton (1952) using rose Bengal. This is powerful protein-specific aqueous stain which colours protoplasm red. Care must be taken to differentiate between shells which absorb stain on the outer surface and those which stain red internally due to the presence of protoplasm. In many cases the protoplasm is stained only in one part, usually in the final chamber. In other cases, the protoplasm may contract so that it does not fill the entire chamber. Lutze (1964) has described examples of *Cribrononion* which have turned green when rose Bengal has been added. They were known to be alive prior to staining because of their pseudopodial activity. Dead foraminiferid shells containing algae or other foreign protoplasm bodies must obviously not be confused with living foraminiferids.

The use of a protoplasmic stain as a means of differentiating living and dead foraminiferids is dependent upon the assumption that shells containing protoplasm were alive at the time of collection. It follows that, upon death, the protoplasm is regarded as being destroyed very quickly. This has been tested experimentally by Boltovskoy and Lena (1970), who found that even ninety-eight days after death, some foraminiferids still had not been decomposed completely. However, bacterial decay is not the only means of destroying protoplasm, and under natural conditions scavengers might speed up the process. But clearly there are situations where freshly dead foraminiferids containing protoplasm could be transported into a depositional environment which differs from the environment of life, e.g. foraminiferids living on seaweeds could be swept up on to a sandy beach. This possibility should be borne in mind when sampling sediments from high-energy environments.

It is customary to county 250 to 300 individuals from each sample to obtain consistent percentage values for the component species. Dryden (1931) has examined the reasons for this, with reference to heavy mineral counts. Ujiié (1962) considers that even with a count of 250 individuals occurrences of less than 7–10 per cent have a sampling error greater than those values and should therefore be rejected. If the standing crop is required, then a portion of sample of known area is counted. If the standing crop is less than 250 individuals, it is desirable that further counts should be made.

The living and dead individuals must be treated separately and, ideally, 250 of each should be counted. Although a number of workers use the total (living plus dead) assemblage data, the author believes this to be wrong. One would not normally include information on the graveyard 'residents' in population statistics on humans because the resulting figures would be meaningless. The same applies to foraminiferids. The living assemblage represents the results of interplay between the fauna and the environment. The dead assemblage reflects short- and long-term changes in the living assemblages, together with postmortem changes. The total assemblage combines all these features, together with a measure of the thickness of the sample (usually 1 cm).

LABORATORY METHODS

The sample is wet-sieved to remove the preservant and the fine-grained sedimentary particles, normally using a 200 (76 μm aperture) or 240 (63 μm aperture) mesh sieve. It is then stained with rose Bengal (1 gram per litre) for $\frac{1}{2}$ to 1 hour, and washed again on a 200 or 240 mesh sieve to remove the surplus stain. After being dried in an oven at 60 °C, it is allowed to cool and then a carbon tetrachloride or trichlorethylene flotation is carried out to concentrate the foraminiferids. The sediment residue should always be checked to ensure that a complete separation of foraminiferids has been made. (See Gibson and Walker (1967) for a discussion of flotation methods.)

The choice of sieve size is important. In practice there is little difference in the results between a 200 and 240 mesh sieve. However, at sieve sizes smaller than 200 mesh, juveniles and small species are lost and consequently the composition of the assemblages is altered.

Appendix 2. Generalized Ecological Data for Selected Genera

The data are given in the order:

1. *Salinity* The known range is given in figures. Where the exact limits are not known the following terms are used—hyposaline (< 33 per mille), normal marine (33–37 per mille), hypersaline (> 37 per mille).

2. *Substrate* Free-living on sediment or clinging to vegetation, pebbles, etc.

3. *Temperature of the bottom water* This is rarely known with precision. Arctic, temperate, subtropical and tropical are used in a geographic sense. They give a rough guide to the bottom temperature.

4. *Depth* The principal range is given. Many species extend into deeper water in very low abundance.

Acervulina Normal marine, attached, ? temperate, 0–60 m, inner shelf.

Alveolinella 39–50 per mille, shell sand, 18–26 °C, 0–6 m, inner shelf and lagoons, Indo-Pacific.

Alveolophragmium Normal marine, muddy sediment, <10 °C, 20–700 m, shelf and upper bathyal.

Ammobaculites Hyposaline, sediment, temperate—? tropical, hyposaline marshes and lagoons and inner shelf.

Ammonia Hyposaline, marine, hypersaline, sediment, 15–30 °C, 0–50 m, hyposaline and hypersaline lagoons, inner shelf.

Ammotium Hyposaline–hypersaline, sediment, 0–30 °C, intertidal—10 m, tidal marshes and hyposaline lagoons, estuaries, and enclosed shelf seas.

Amphistegina >34 per mille, sea-grass, sediment, coral reefs, 25–26 °C, 5–20 m, inner shelf.

Archaias Normal marine—hypersaline, ? sediment and vegetation, tropical, inner shelf.

Arenoparrella Hyposaline–hypersaline, sediment, 0–30 °C, intertidal, tidal marshes. Some species known from 82–460 m off California.

Asterigerina Normal marine, sediment, tropical–subtropical, inner shelf.

Asterigerinata 35 per mille, sediment, temperate, 0–100 m, inner shelf.

Baculogypsina Normal marine, algae, coral fragments, tropical >25 °C, intertidal to ? 10 m, coral reefs.

Bigenerina Normal marine, muddy sediment, cold, >100 m, outer shelf and bathyal.

Bolivina   32–36 per mille, muddy sediment, 1–30 °C, bathyal to marginal marine.

Borelis   Probably as for *Alveolinella*, but also Atlantic.

Brizalina   32–36 per mille, muddy sediment, 1–30 °C, bathyal to marginal marine.

Buccella   Normal marine, muddy sediment, cold to warm temperate, 0–180 m, shelf.

Bulimina   32–36 per mille, muddy sediment, 1–30 °C, bathyal to nearshore.

Buliminella   Normal marine, muddy sediment, temperate, 0–800 m, mainly shelf but extends into bathyal.

Calcarina   Normal marine, algae, coral fragments, tropical >25 °C, intertidal to ? 10 m, coral reefs.

Cancris   Normal marine, sediment, temperate–subtropical, 50–150 m, shelf.

Cassidulina   Normal marine, muddy sediment, cold to temperate, 5–>3000 m, shelf to bathyal.

Chilostomella   Normal marine, muddy sediment, cold, 80–1190 m, outer shelf and bathyal.

Cibicides   Normal marine, clinging to vegetation, stones, shells, living animals, arctic to tropical, 0–>2000 m, shelf to bathyal.

Cribrostomoides   Slightly hyposaline (30 per mille) to normal marine, sandy sediments, ?<15 °C, 0–150 m, shelf.

Cyclammina   Normal marine, sediment, 10 °C, sigma – *t* 27.7, >100 m, outer shelf and upper bathyal.

Cyclogyra   Normal marine, sediment, cold to tropical, 0–100 m, inner shelf.

Discorbis   Normal marine, vegetation, >12 °C, 0–50 m, inner shelf.

Eggerella   20–37 per mille, sediment, arctic to temperate, 0–100 m, inner shelf, enclosed hyposaline shelf-seas and lagoons.

Elphidium   Keeled: 35–50 per mille, sediment and vegetation, >15 °C, 0–50 m, inner shelf.
Unkeeled: 0–70 per mille, sediment and vegetation, 1–30 °C, 0–50 m, hyposaline to hypersaline tidal marshes and lagoons, nearshore.

Eponides   Normal marine, sediment, cold to temperate, 10–>6000 m, shelf to bathyal.

Fissurina   Normal marine, muddy sediment, cold to tropical, 0–150 m, shelf.

Fursenkoina   Slightly hyposaline to normal marine, muddy sediment, temperate, 0–1190 m, lagoons, shelf and bathyal.

Gaudryina   Normal marine, sandy sediment, temperate, 50–460 m, shelf and upper bathyal.

Gavelinopsis   Normal marine, clinging to vegetation, stones and shells, temperate, 0–>1000 m, shelf and bathyal.

Glabratella   >35 per mille, sediment or vegetation, temperate–tropical, 0–50 m, hypersaline tidal marshes and lagoons, normal marine inner shelf.

Globobulimina   Slightly hyposaline (32 per mille) to normal marine, muddy sediment, <10 °C, 20–2000 m, shelf to bathyal.

Globulina   Normal marine, sediment, temperate to tropical, 0–60 m, inner shelf.

Gyroidina   Normal marine, muddy sediment, cold, shelf and bathyal.

Heterostegina   Normal marine to hypersaline, ? sediment or ? vegetation, tropical, inner shelf.

Hoeglundina   Normal marine, muddy sediment, <5 °C, 140—1100 m, outer shelf and bathyal.

Hyalinea   Normal marine, muddy sediment, cold to temperate, 10–1000 m, shelf and bathyal.

Islandiella   32–35 per mille, muddy sediment, <10 °C, >20 m, shelf.

Jadammina   0–50 per mille, sediment, 0–30 °C, intertidal, tidal marshes.

Karreriella   Normal marine, muddy sediment, <10 °C, >100 m, outer shelf and upper bathyal.

Lagena   Normal marine, muddy sediment, cold to tropical, 0–180 m, shelf.

Lenticulina   Normal marine, muddy sediment, cold, >100 m, outer shelf and bathyal.

Marginopora   Normal marine to hypersaline, sea-grass and seaweed, 18–26 °C, 0–8 m, inner shelf.

Martinottiella   Normal marine, muddy sediment, cold, >120 m, outer shelf and bathyal.

Massilina   Normal marine, vegetation and sandy sediment, temperate to subtropical, 0–40 m, inner shelf.

Melonis   Normal marine, muddy sediment, <10 °C, 5–1000 m, shelf and bathyal.

Miliammina   0–50 per mille, sediment, 0–30 °C, 0–10 m, hyposaline lagoons, hyposaline—hypersaline tidal marshes.

Miliolinella   32–50 per mille, sediment, 10–30 °C, 0–100 m, inner shelf, normal marine lagoons and tidal marshes, hypersaline lagoons.

Nonion   Hyposaline to normal marine, sediment, cold to tropical, 0–180 m, shelf.

Nonionella   Normal marine, muddy sediment, temperate—subtropical, 10–1000 m, shelf and bathyal.

Oolina   Normal marine, muddy sediment, cold to tropical, 0–180 m, shelf.

Operculina   Normal marine—hypersaline, sediment, tropical, ? inner shelf.

Paromalina   Normal marine, muddy sediment, <10 °C, >150 m, outer shelf and bathyal.

Parrina   Hypersaline, sediment and vegetation, 15–30 °C, 0–20 m, lagoons and nearshore.

Patellina   Normal marine, clinging to pebbles or shells, cold to subtropical, 0–100 m, inner shelf.

Peneroplis   37–53 per mille, sea-grass, seaweed, algae, 18–27 °C, 0–35 m, normally 0–10 m, lagoons and nearshore.

Planorbulina   Normal marine, attached to vegetation, stones and shells, temperate—subtropical, 0–50 m, inner shelf.

Protelphidium   0–35 per mille, sediment, 0–20 °C, 0–10 m, hyposaline tidal marshes, lagoons and estuaries.

Pullenia   Normal marine, muddy sediment, <10 °C, 150–6000 m, outer shelf to deep sea.

Pyrgo   Group 1: normal marine, vegetation and sediment, temperate, 0–50 m, inner shelf.

Group 2: normal marine, muddy sediment, <5 °C, 50–2000 m, shelf and bathyal.

Quinqueloculina >32 per mille, sandy sediment and vegetation, mainly temperate to tropical but some arctic species, 0–40 m, inner shelf, normal marine and hypersaline lagoons.

Reophax Normal marine, sediment, arctic—tropical, 0–150 m, shelf.

Rosalina Normal marine, clinging to vegetation, shells, stones, temperate—subtropical, 0–100 m, inner shelf.

Rotalia 36–38 per mille, sediment, 14–25 °C, 0–40 m, inner shelf.

Saccammina 32–36 per mille, sandy sediment, <15 °C, 0–100 m, inner shelf.

Siphotextularia Normal marine, muddy sediment, ?<10 °C, 150–>1000 m outer shelf and bathyal.

Sorites >37 per mille, sea-grass, 18–26 °C, 0–35 m, lagoons and near-shore.

Spirillina Normal marine, clinging to stones and shells, cold to temperate, 0–100 m, inner shelf.

Spirolina >37 per mille, sea-grass, 18–26 °C, 0–35 m, lagoons and near-shore.

Stetsonia Normal marine, muddy sediment, temperate to subtropical, 0–60 m, inner shelf.

Textularia Normal marine, sandy sediment, arctic to tropical, 50–640 m, shelf and upper bathyal.

Trifarina Normal marine, muddy sediment, temperate, 0–400 m, shelf and upper bathyal.

Triloculina >32 per mille, sandy sediment and vegetation, mainly temperate to tropical, 0–40 m, inner shelf, normal marine, and hypersaline lagoons. Some bathyal species.

Trochammina Group 1: hyposaline to hypersaline, muddy sediment, 0–30 °C, intertidal, tidal marshes.
Group 2: normal marine, sediment, cold to temperate, 0–2000 m, shelf and bathyal.

Uvigerina Normal marine, muddy sediment, cold, 100–>4500 m, outer shelf to bathyal.

Vertebralina Normal marine to hypersaline, vegetation, 15–30 °C, 0–20 m, lagoons and inner shelf.

Appendix 3. Problems Requiring Further Study

1. Most studies have been concerned with intertidal, lagoonal and shallow-water nearshore areas. Much more distributional and assemblage data are needed from shelf, slope and deep seas, coral reefs, and all tropical and arctic shallow-water environments.

2. Long-term studies of small areas are necessary to attempt to relate the annual and diurnal variations in as many environmental parameters as possible to changes in the abundance and composition of foraminiferid assemblages. Ideally, such studies should be carried out in shelf seas.

3. Studies of standing crop size and annual production are necessary to assess the fertility of the sea floor and to aid the understanding of the relationship between the living and dead assemblages.

4. The position of foraminiferids in marine communities is an open field for investigation and should produce valuable results both for biologists and geologists.

5. It is probable that the microspheric and megalospheric generations have different ecological requirements. Nothing is known of this at present.

6. The enormous variety of structure, form and ornament shown by the tests of foraminiferids provides ample scope for studies of evolution and classification. However, little is known of the functional significance of these features or of the environmental influence on them.

7. Pollution studies already carried out show that sewerage changes the composition of foraminiferid assemblages and influences the abundance of individuals. This important field needs much further study.

References

ANGELL, R. W. 1967 The process of chamber formation in the foraminifer *Rosalina floridana* (Cushman). *J. Protozool.*, 14, 566–74.

ARNOLD, Z. M. 1953 An introduction to the study of movement and dispersal in *Allogromia laticollaris* Arnold. *Contr. Cushman Fdn foramin. Res.*, 4, 15–21.

ARNOLD, Z. M. 1954 Culture methods in the study of living Foraminifera. *J. Paleont.*, 28, 404–16.

ARNOLD, Z. M. 1966 A laboratory system for maintaining small-volume cultures of Foraminifera and other organisms. *Micropaleontology*, 12, 109–18.

ARNOLD, Z. M. 1967 Biological clues in interpreting the biogeography of the foraminifer *Nubecularia lucifuga* Defrance. *Univ. Miami: Studies in Tropical Oceanography*, no. 5, 622–31.

ATKINSON, K. 1969 The association of living Foraminifera with algae from the littoral zone, south Cardigan Bay, Wales. *J. nat. Hist.*, 3, 517–42.

AYALA-CASTAÑARES, A. 1963 Sistematica y distribucion de los Foraminiferos recientes de la Laguna de Terminos, Campeche, Mexico. *Boln Inst. Geol. Méx.*, 67, 1–130.

AYALA-CASTAÑARES, A. and SEGURA, L. R. 1968 Ecologia y distribucion de los Foraminiferos recientes de la Laguna Madre, Tamaulipas, Mexico. *Boln. Inst. Geol. Méx.*, 87, 1–89.

BANDY, O. L. 1956 Ecology of Foraminifera in northeastern Gulf of Mexico. *Prof. Pap. U.S. geol. Surv.*, 274 G, 179–204.

BANDY, O. L. 1963 Larger living Foraminifera of the continental borderland of southern California. *Contr. Cushman Fdn foramin. Res.*, 14, 121–6.

BANDY, O. L. 1964 General correlation of foraminiferal structure with environment, In: Imbrie, J. and Newell, D., *Approaches to paleoecology*, John Wiley and Sons, New York, 75–90.

BANDY, O. L., INGLE, J. C. and RESIG, J. M. 1964a Foraminifera, Los Angeles County outfall area, California. *Limnol. Oceanogr.*, 9, 124–37.

BANDY, O. L., INGLE, J. C. and RESIG, J. M. 1964b Foraminiferal trends, Laguna Beach outfall area, California. *Limnol. Oceanogr.*, 9, 112–23.

BANDY, O. L., INGLE, J. C. and RESIG, J. M. 1964c Facies trends, San Pedro Bay, California. *Bull. geol. Soc. Am.*, 75, 403–24.

BANDY, O. L., INGLE, J. C. and RESIG, J. M. 1965a Foraminiferal trends, Hyperion outfall, California. *Limnol. Oceanogr.*, 10, 314–32.

BANDY, O. L., INGLE, J. C. and RESIG, J. M. 1965b Modification of foraminiferal distribution by the Orange County outfall, California. *Ocean Science and Ocean Engineering*, 55–76.

BANDY, O. L. and RODOLFO, K. S. 1964 Distribution of Foraminifera and sediments, Peru-Chile Trench area. *Deep-Sea Res.*, 11, 817–37.

BANNER, F. T. 1970 A synopsis of the Spirocyclinidae. *Revista Española de Micropaleontologia*, 2, 243–90.

BANNER, F. T. 1971 A new genus of the Planorbulinidae an endoparasite of another foraminifer. *Revista Española de Micropaleontologia*, 3, 113–28.

BARTLETT, G. A. 1964 M.S.a. Benthonic foraminiferal ecology in St. Margarets Bay and Mahone Bay, south-east Nova Scotia. Bedford Inst. Oceanogr., Rept. B.I.O. 64–8, unpublished M.S.

BARTLETT, G. A. 1965 M.S.b. Preliminary investigation of benthic foraminiferal ecology in Tracadie Bay, Prince Edward Island. Bedford Inst. Oceanogr., Rept. B.I.O. 65–3, unpublished M.S.

BARTLETT, G. A. 1966 M.S. Distribution and abundance of Foraminifera and Thecamoebina in Miramichi River and Bay. Bedford Inst. Oceanogr., Rept. B.I.O., 66–2, unpublished M.S.

BEERBOWER, J. R. and JORDAN, D. 1969 Application of information theory to paleontologic problems: taxonomic diversity. *J. Paleont.*, **43**, 1184–98.

BENSON, R. H. 1959 Ecology of recent ostracodes of the Todos Santos Bay region, Baja California, Mexico. *Paleont. Contr. Univ. Kans.*, **23**, 1–80.

BIGELOW, H. B. 1933 Studies of the waters of the continental shelf, Cape Cod to Chesapeake Bay. 1. the cycle of temperature. *Pap. phys. Oceanogr. Met.*, **2**, 1–135.

BLANC-VERNET, L. 1965 Note sur la répartition des Foraminifères au voisinage des côtes de Terre Adélie (Antarctique). *Recl. Trav. Stn mar. Endoume*, **36**, 191–206.

BLANC-VERNET, L. 1969 Contribution à l'étude des Foraminifères de Méditerranée. *Recl. Trav. Stn mar. Endoume*, **48**, 5–281.

BOCH, W. D. 1970 *Thalassia testudinarium*, a habitat and means of dispersal for shallow water benthonic foraminifera. *Trans. Gulf-Cst Ass. geol. Socs*, **19**, 337–40.

BOLTOVSKOY, E. 1957 Las anormalidades en las caparazones de Foraminiferos y el 'Indice de regeneramiento'. *Ameghiniana*, **1**, 80–4.

BOLTOVSKOY, E. 1963a The littoral foraminiferal biocoenoses of Puerto Deseado (Patagonia, Argentina). *Contr. Cushman Fdn foramin. Res.*, **14**, 58–70.

BOLTOVSKOY, E. 1963b Sobre las relaciones entre Foraminiferos y turbelarios. *Neotropica*, **9**, 55–60.

BOLTOVSKOY, E. 1963c Foraminiferos y sus relaciones con el medio. *Revta Mus. argent. Cienc. nat. Bernardino Rivadavia Inst. nac. Invest. Cienc. nat.*, **1**, 21–107.

BOLTOVSKOY, E. 1964 Seasonal occurrences of some living Foraminifera in Puerto Deseado (Patagonia, Argentina). *J. Cons. perm. int. Explor. Mer*, **39**, 136–45.

BOLTOVSKOY, E. 1966 Depth at which Foraminifera can survive in sediments. *Contr. Cushman Fdn foramin. Res.*, **17**, 43–5.

BOLTOVSKOY, E. 1970 Distribution of marine littoral Foraminifera in Argentina, Uruguay and Southern Brazil. *Mar. Biol.*, **6**, 335–44.

BOLTOVSKOY, E. and LENA, H. 1969a Microdistribution des Foraminifères benthoniques vivants. *Revue Micropaléont.*, **12**, 177–85.

BOLTOVSKOY, E. and LENA, H. 1969b Seasonal occurrences, standing crop and production in benthic foraminifera of Puerto Deseado. *Contr. Cushman Fdn foramin. Res.*, **20**, 87–95.

BOLTOVSKOY, E. and LENA, H. 1969c Los epibiontes de *Macrocystis* flotante como indicadores hidrologicos. *Neotropica*, **15**, 135–7.

BOLTOVSKOY, E. and LENA, H. 1970. On the decomposition of the protoplasm and the sinking velocity of the planktonic foraminifers. *Int. Revue ges. Hydrobiol.*, **55**, 797–804.

BOLTOVSKOY, E. and LENA, H. 1971 The Foraminifera (except Family Allogromidae) which dwell in fresh water. *J. foramin. Res.*, **1**, 71–6.

BOLTOVSKOY, E. and THEYER, F. 1970 Foraminferos recientes de Chili Central. *Revta Mus. argent. Cienc. nat. Bernardino Rivadavia Inst., Hidrobiol.*, **2**, 279–380.

BRADSHAW, J. S. 1955 Preliminary laboratory experiments on ecology of foraminiferal populations. *Micropaleontology*, **1**, 351–8.

BRADSHAW, J. S. 1957 Laboratory studies on the rate of growth of the foraminifer 'Streblus beccarii (Linné) var. tepida (Cushman)'. *J. Paleont.*, **31**, 1138–47.

BRADSHAW, J. S. 1961 Laboratory experiments on the ecology of foraminifera. *Contr. Cushman Fdn foramin. Res.*, **12**, 87–106.

BRADSHAW, J. S. 1968 Environmental parameters and marsh Foraminifera. *Limnol. Oceanogr.*, **13**, 26–38.

BRADY, H. B. 1884 Report on the Foraminifera dredged by H.M.S. Challenger during the years 1873–1876. *Rept. Voy. Challenger*, Zool., **9**, 1–814.

BROOKS, A. L. 1967 Standing crop, vertical distribution, and morphometrics of *Ammonia beccarii* (Linné). *Limnol. Oceanogr.*, **12**, 667–84.

BUCHANAN, J. B. and HEDLEY, R. H. 1960 A contribution to the biology of *Astrorhiza limicola* (Foraminifera). *J. mar. biol. Ass. U.K.*, **39**, 549–60.

BUZAS, M. A. 1965 The distribution and abundance of Foraminifera in Long Island Sound. *Smithson. misc. Collns*, **149**, 1–89.

BUZAS, M. A. 1967 An application of canonical analysis as a method of comparing faunal areas. *J. Anim. Ecol.*, **36**, 563–77.

BUZAS, M. A. 1968a On the spatial distribution of Foraminifera. *Contr. Cushman Fdn foramin. Res.*, **19**, 1–11.

BUZAS, M. A. 1968b Foraminifera from the Hadley Harbor Complex, Massachusetts. *Smithson. misc. Collns*, **152**, (8) 1–26.

BUZAS, M. A. 1969 Foraminiferal species densities and environmental variables in an estuary. *Limnol. Oceanogr.*, **14**, 411–22.

BUZAS, M. A. 1970 Spatial homogeneity: statistical analyses of unispecies and multispecies populations of Foraminifera. *Ecology*, **51**, 874–9.

BUZAS, M. A. and GIBSON, T. G. 1969 Species diversity: benthonic Foraminifera in Western North Atlantic. *Science*, N.Y., **163**, 72–5.

CARRIGY, M. A. 1956 Organic sedimentation in Warnboro Sound, Western Australia. *J. sedim. Petrol.*, **26**, 228–39.

CEBULSKI, D. E. 1961 Distribution of Foraminifera in the barrier reef and lagoon of British Honduras. Texas A. and M. Research Foundation, Project 24, reference 61–13T. Unpublished M.S. privately circulated.

CEBULSKI, D. E. 1962 Foraminiferal populations and faunas in the barrier reef and lagoon of British Honduras. *Trans. Gulf-Cst Ass. geol. Socs.*, **12**, 283–4.

CEBULSKI, D. E. 1969 Foraminiferal populations and faunas in barrier-reef tract and lagoon, British Honduras. *Mem. Am. Ass. Petrol. Geol.*, **11**, 311–28.

CHIJI, M. and LOPEZ, S. M. 1968 Regional foraminiferal assemblages in Tanabe Bay, Kii Peninsula, Central Japan. *Publ. Seto mar. biol. Lab.*, **16**, 85–125.

CHRISTIANSEN, B. O. 1964 *Spiculosiphon radiata*, a new Foraminifera from Northern Norway. *Astarte*, **25**, 1–8.

CIVRIEUX, J. M. S. de 1968 Un metodo de observaciones y representaciones graficas del metabolismo (foraminiferos) con aplicacion al estudio de la ecologia en biotopos de salinidad altamente variable. *Bol. Inst. Oceanog.*, Univ. Oriente, **7**, 99–109.

CLOSS, D. 1963 Foraminiferos e Tecamebas da Lagoa dos Patos (R.G.S.) *Bolm Esc. Geol. Univ. Pôrto Alegre*, **11**, 1–130.

CLOSS, D. and MADEIRA, M. 1962 Tecamebas e Foraminiferos do Arroio Chui (Santa Vitoria do Palmar, R. Grande do Sul, Brasil). *Iheringia*, Zool., **19**, 3–43.

CLOSS, D. and MADEIRA, M. 1966 Foraminifera from the Paranagua Bay, State of Parana, Brazil. *Bolm Univ. Parana, Conselho de Pesquisas*, Zool., **2**, 139–62.

CLOSS, D. and MADEIRA, M. 1967 Foraminiferos e Tecamebas aglutinantes da Lagoa de Tremendai, No Rio Grande do Sul. *Iheringia*, Zool., **35**, 7–31.

CLOSS, D. and MADEIRA, M. L. 1968 Seasonal variations of brackish Foraminifera in the Patos Lagoon, Southern Brazil. *Esc. Geol. Porto Allegre Publ. Esp.*, **15**, 1–51.

CLOSS, D. and MADEIROS, V. M. F. 1965 New observations on the ecological subdivisions of the Patos Lagoon in Southern Brazil. *Bol. Instituto Ciencias Naturais, Rio Grande do Sul*, 24, 7–33.

CLOSS, D. and MADEIROS, V. M. F. 1967 Thecamoebina and Foraminifera from the Mirim Lagoon, southern Brazil. *Iheringia*, Zool., 35, 75–88.

COOPER, L. H. N. 1966 English Channel. In: Fairbridge, R. W., Ed., *The Encyclopedia of Oceanography, Encyclopedia of Earth Science Series*, vol. 1, Reinhold Publishing Corporation, New York, 256–64.

COOPER, L. H. N. 1967 The physical oceanography of the Celtic Sea. *Oceanography and Marine Biology, an annual review*, 5, 99–110.

COOPER, W. C. 1961 Intertidal Foraminifera of the California and Oregon coast. *Contr. Cushman Fdn foramin. Res.*, 12, 47–63.

CUSHMAN, J. A. 1922 Shallow water Foraminifera of the Tortugas region. *Pap. Tortugas Lab.*, 17, 1–85.

DANIELS, C. H. 1970 Quantitative ökologische Analyse der zeitlichen und räumlichen verteilung rezenter Foraminiferen im Limski Kanal bei Rovinj (nördliche Adria). *Göttinger Arb. geol. Paläont.*, 8, 1–109.

DAVIES, G. R. 1970 Carbonate bank sedimentation, Eastern Shark Bay, Western Australia. *Mem. Am. Ass. Petrol. Geol.*, 13, 85–168.

DENNISON, J. M. and HAY, W. H. 1967 Estimating the needed sampling area for subaquatic ecologic studies. *J. Paleont.*, 41, 706–8.

DOYLE, W. L. and DOYLE, M. M. 1940 The structure of Zooxanthellae. *Pap. Tortugas Lab.*, 32, 127–42.

DRYDEN, A. L. 1931 Accuracy in percentage representation of heavy mineral frequencies. *Proc. natn Acad. Sci. U.S.A.*, 17, 233–8.

EHRENBERG, C. G. 1839 Ueber noch jetzt zahlreich lebende Thierarten der Kreidebildung und der Organismus der Polythalamien. *Abh. Akad. Wiss. Berlin*.

EKMAN, S. 1953. *Zoogeography of the Sea*. Sidgwick and Jackson, London.

ELLISON, R. L. and NICHOLS, M. M. 1970 Estuarine Foraminifera from the Rappahannock River, Virginia. *Contr. Cushman Fdn foramin. Res.*, 21, 1–17.

EMERY, K. O. 1956 Sediments and water of Persian Gulf. *Bull. Am. Ass. Petrol. Geol.*, 40, 2354–83.

EMERY, K. O. 1960 *The sea off southern California*. John Wiley and Sons Inc., New York, 366 pp.

EMERY, K. O. 1968 Relict sediments on continental shelves of World. *Bull. Am. Ass. Petrol. Geol.*, 52, 445–64.

EMERY, K. O., BUTCHER, W. S., GOULD, H. R. and SHEPARD, F. P. 1952 Submarine geology off San Diego, California. *J. Geol.*, 60, 511–48.

EMERY, K. O. and STEVENSON, R. E. 1957 Estuaries and lagoons. *Mem. geol. Soc. Am.*, 67, (1), 673–750.

EVANS, G. 1965 Intertidal flat sediments and their environments of deposition in the Wash. *Q. Jl geol. Soc. Lond.*, 121, 209–45.

EVANS, G. 1966a The recent sedimentary facies of the Persian Gulf region. *Phil. Trans. R. Soc.*, A, 259, 291–8.

EVANS, G. 1966b Persian Gulf. In: Fairbridge, R. W., ed., *The Encyclopedia of Oceanography, Encyclopedia of the Earth Science Series*, vol. 1, Reinhold Publishing Corporation, New York, 689–95.

EVANS, G. and BUSH, P. 1969 Some oceanographical and sedimentological observations on a Persian Gulf lagoon. *Mem. Simp. Intern. Lagunas Costeras*, UNAM-UNESCO, 155–70.

EVANS, G., KINSMAN, D. J. J., and SHEARMAN, D. J. 1964 A reconnaissance survey of the environment of Recent carbonate sedimentation along the Trucial Coast, Persian Gulf. *Developments in Sedimentology*, **1**, 129–35.

FISHER, P. H. 1966. Ecologie de certains Foraminifères du haut niveau sur les côtes du Pacifique. *Bull. Soc. Zool. Fr.*, **91**, 295–300.

FISHER, R. A., CORBETT, A. S. and WILLIAMS, C. B. 1943 The relationship between the number of species and the number of individuals in a random sample of an animal population. *J. Anim. Ecol.*, **12**, 42–58.

FORTI, I. R. S. and ROETTGER, E. 1967 Further observations on the seasonal variations of mixohaline Foraminifera from the Patos Lagoon, southern Brazil. *Arch. Oceanogr. Limnol.*, **15**, 55–61.

FØYN, B. 1934 Lebenszyklus, Cytologie und Sexualität der Chlorophycee *Cladophora suriana* Kützing. *Arch. Protistenk.*, **83**, 1–56.

FRANKEL, L. 1972 Subsurface reproduction in Foraminifera. *J. Paleont.*, **46**, 62–5.

FREUDENTHAL, H. D., LEE, J. J. and PIERCE, S. 1963 Growth and physiology of Foraminifera in the laboratory: Part 2—a tidal system for laboratory studies on eulittoral foraminifera. *Micropaleontology*, **9**, 443–8.

FUNNELL, B. M. 1967 Foraminifera and Radiolaria as depth indicators in the marine environment. *Mar. geol.*, **5**, 333–47.

GIBSON, L. B. 1966 Some unifying characteristics of species diversity. *Contr. Cushman Fdn foramin. Res.*, **17**, 117–24.

GIBSON, T. G. and WALKER, W. 1967 Flotation methods for obtaining Foraminifera from sediment samples. *J. Paleont.*, **41**, 1294–7.

GLAESSNER, M. F. 1963 Major trends in the evolution of the Foraminifera. In: Von Koenigswald, G. H. R., Emeis, J. D., Buning, W. L. and Wagner, C. W., eds., *Evolutionary Trends in Foraminifera*, Elsevier Publishing Company, Amsterdam, 9–24.

GRABERT, B. 1971 Zur Eignung von Foraminiferen als Indikatoren für Sandwanderung. *Sonderdr. Deutsch. Hydr. Zeitschr.*, **24**, 1–14.

GRAHAM, A. 1955 Molluscan diets. *Proc. malac. Soc. Lond.*, **31**, 144–59.

GREEN, K. E. 1960 Ecology of some Arctic Foraminifera. *Micropaleontology*, **6**, 57–78.

HAAKE, W. H. 1962 Untersuchungen an der Foraminiferen-fauna in Wattgebiet zwischen Langeoog und dem Festland. *Meyniana*, **12**, 25–64.

HAAKE, F. W. 1967 Zum Jahresgang von Populationen einer Foraminiferen-Art in der westlichen Ostsee. *Meyniana*, **17**, 13–27.

HAAKE, F. W. 1970 Zur Tiefenverteilung von Miliolinen (Foram.) im Persischen Golf. *Paläont. Z.*, **44**, 196–200.

HAMAN, D. 1969 Seasonal occurrence of *Elphidium excavatum* (Terquem) in Llandanwg Lagoon (North Wales, U.K.) *Contr. Cushman Fdn foramin. Res.*, **20**, 139–42.

HAMAN, D. 1972 Foraminiferal assemblages in Tremadoc Bay, North Wales, U.K. *J. Foramin. Res.*, **1**, 126–43 (for 1971).

HANSEN, H. J. 1972 Pore pseudopodia and sieve plates of *Amphistegina*. *Micropaleontology*, **18**, 223–30.

HAYNES, J. 1965 Symbiosis, wall structure and habitat in Foraminifera. *Contr. Cushman Fdn foramin. Res.*, **16**, 40–3.

HAYNES, J. and DOBSON, M. 1969 Physiography, foraminifera and sedimentation in the Dovey Estuary (Wales). *Geol. J.*, **6**, 217–56.

HEDGPETH, J. W. 1957a Classification of marine environments. *Mem. geol. Soc. Am.*, **67**, (1), 17–28.

HEDGPETH, J. W. 1957b Estuaries and lagoons II Biological aspects. *Mem. geol. Soc. Am.*, **67**, (1), 693–750.

HEDLEY, R. H. 1958 A contribution to the biology and cytology of *Haliphysema* (Foraminifera). *Proc. zool. Soc. Lond.*, **130**, 569–76.

HEDLEY, R. H. 1964 The biology of Foraminifera. In: Felts, W. J. L. and Harrison, R. J. 1964 *Int. Rev. gen. exp. Zool.*, **1**, 1–45.

HEDLEY, R. H., HURDLE, C. M. and BURDETT, I. D. J. 1967 The marine fauna of New Zealand: intertidal Foraminifera of the *Corallina officinalis* zone, *Mem. N.Z. oceanogr. Inst.*, **38**, 1–86.

HEDLEY, R. H. and WAKEFIELD, J. ST. J. 1967 Clone culture of a new rosalinid foraminifer from Plymouth, England and Wellington, New Zealand. *J. mar. biol. Ass. U.K.*, **47**, 121–8.

HERON-ALLEN, E. 1915 Contributions to the study of the bionomics and reproductive processes of the foraminifera. *Phil. Trans. R. Soc.*, B, **206**, 227–79.

HERON-ALLEN, E. and EARLAND, A. 1910 On the recent and fossil Foraminifera of the shore-sands of Selsey Bill, Sussex. VI Acontribution towards the aetiology of *Massilina secans* (d'Orbigny sp.). *Jl. R. microsc. Soc.*, 693–5.

HESSLER, R. R. and SANDERS, H. L. 1967 Faunal diversity in the deep-sea. *Deep-Sea Res.*, **14**, 65–78.

HILTERMANN, H. 1948 Klassifikation der natürlichen Brackwässer. *Erdöl Kohle*, **2**, 4–8.

HOFKER, J. 1931 De Foraminiferen in den omtrek van Amsterdam. De Biologie van der Zuidersee tijdens haar drooglegging. Afl. 3, 61–6.

HOLME, N. A. 1964 Methods of sampling benthos. *Adv. mar. Biol.*, **2**, 171–260.

HOPKINS, T. L. 1964 A survey of marine bottom samplers. *Progress in Oceanography*, **2**, 213–56.

HOWARD, J. F., KISSLING, D. L. and LINEBACK, J. A. 1970 Sedimentary facies and distribution of biota in Coupon Bight, Lower Florida Keys. *Bull. geol. Soc. Am.*, **81**, 1929–46.

HOWARTH, R. J. and MURRAY, J. W. 1969 The Foraminiferida of Christchurch Harbour, England: a reappraisal using multivariate techniques, *J. Paleont.*, **43**, 660–75.

IKEYA, N. 1970 Population ecology of benthonic Foraminifera in Ishikari Bay, Hokkaido, Japan. *Rec. oceanogr. Wks Japan*, **10**, 173–91.

IKEYA, N. 1971 Species diversity of benthonic Foraminifera, off the Shimokita Peninsula, Pacific Coast of North Japan. *Rec. oceanogr. Wks Japan*, **11**, 27–37.

JELL, J. S., MAXWELL, W. H. G. and MCKELLAR, R. G. 1965 The significance of the larger Foraminifera in the Heron Island reef sediments. *J. Paleont.*, **39**, 273–9.

JEPPS, M. W. 1942 Studies on *Polystomella* Lamarck (Foraminifera). *J. mar. biol. Ass. U.K.*, **25**, 607–66.

JEPPS, M. W. 1956 *The Protozoa, Sarcodina*. Oliver and Boyd, Edinburgh. 183 pp.

KAESLER, R. L. 1969 Quantitative re-evaluation of ecology and distribution of recent Foraminifera and Ostracoda of Todos Santos Bay, Baja California, Mexico. *Paleont. Contr. Univ. Kans.*, **10**, 1–50.

KETCHUM, B. H. and CORWIN, N. 1964 The persistence of 'winter' water on the continental shelf south of Long Island, New York. *Limnol. Oceanogr.*, **9**, 467–75.

KRUIT, C. 1955 *Sediments of the Rhône Delta: Part 1 Grain size and microfauna*. Mounton and Co., S-Gravenhage, 1–141.

LANKFORD, R. R. 1959 Distribution and ecology of Foraminifera from East Mississippi Delta margin. *Bull. Am. Ass. Petrol. Geol.*, **43**, 2068–99.

LAUFF, G. H. 1967 Estuaries. *Publs. Am Ass. Advmt Sci.*, **83**, 1–757.

LE CALVEZ, J. 1938 Un Foraminifère géant *Bathysiphon filiformis* G.O. Sars. *Arch. Zool. Exp. et Gén.*, **79**, 82–8.

LE CALVEZ, J. 1940 Une amibe, *Wahlkampfia discorbini* n. sp., parasite du foraminifère *Discorbis mediterranensis* (d'Orbigny). *Arch. Zool. Exp. et Gén.*, 81, 123–9.

LE CALVEZ, J. 1947 *Entosolenia marginata* (Walker and Boys) Foraminifère apogamique ectoparasite d'une autre foraminifère: *Discorbis villardeboanus*(d'Orbigny). *C. r. hebd. Scéanc. Acad. Sci.*, Paris, 224, 1448–50.

LE CALVEZ, J. and LE CALVEZ, Y. 1958 Répartition des Foraminifères dans la Baie de Villefranche 1—Miliolidae. *Annls Inst. océanogr.*, Monaco, 35, 159–234.

LE CAMPION, J. 1970 Contribution à l'étude des Foraminifères du Bassin d'Arcachon et du proche océan. *Bull. Inst. Géol. Bassin Aquitaine*, 8, 3–98.

LEE, J. J. and FREUDENTHAL, H. 1964 Neglected Amoebas in culture. *Natural History Magazine*, Am. Mus. nat. Hist., 54–61.

LEE, J. J., FREUDENTHAL, H. D., MULLER, W. A., KOSSOY, V., PIERCE, S. and GROSSMAN, R. 1963 Growth and physiology of Foraminifera in the laboratory: Part 3—initial studies on *Rosalina floridana* (Cushman). *Micropaleontology*, 9, 449–66.

LEE, J. J. and MARCELLINO, C. L. 1967 The effect of selected pollutants on *Allogromia laticollaris*. *J. Protozool.*, Suppl., 14, 16, Abstract 46.

LEE, J. J., MCENERY, M., PIERCE, S., FREUDENTHAL, H. D. and MULLER, W. A. 1966 Tracer experiments in feeding littoral foraminifera. *J. Protozool.*, 13, 659–70.

LEE, J. J., MCENERY, M., PIERCE, S., and MULLER W. A. 1966 Prey and predator relationships in the nutrition of certain littoral foraminifera. *J. Protozool.*, Suppl., 13, 23, Abstract 86.

LEE, J. J. and MULLER, W. A. 1967 Growth rates of Foraminifera in gnotobiotic culture. *J. Protozool.*, Suppl., 14, 19, Abstract 60.

LEE, J. J., MULLER, W. A., STONE, R. J., MCENERY, M. E. and ZUCKER, W. 1969 Standing crop of Foraminifera in sublittoral epiphytic communities of a Long Island salt marsh. *Mar. Biol.*, 4, 44–61.

LEE, J. J. and PIERCE, S. 1963 Growth and physiology of Foraminifera in the laboratory: Part 4—monoxenic culture of an Allogromiid with notes on its morphology. *J. Protozool.*, 10, 401–11.

LEE, J. J., PIERCE, S., TENTCHOFF, M. and MCLAUGHLIN, J. J. 1961 Growth and physiology of Foraminifera in the laboratory: Part 1—collection and maintenance. *Micropaleontology*, 7, 461–6.

LEE, J. J. and ZUCKER, W. 1969 Algal flagellate symbiosis in the foraminifer *Archaias*. *J. Protozool.*, 16, 71–81.

LEES, A., BULLER, A. T. and SCOTT, J. 1969 Marine carbonate sedimentation processes, Connemara, Ireland. Reading University Geological Reports, no. 2, Unpublished M.S.

LENA, H. 1966 Foraminiferos recientes de Ushuaia (Tierra del Fuego, Argentina). *Ameghiniana*, 4, 311–36.

LESLIE, R. J. 1965 Ecology and paleoecology of Hudson Bay Foraminifera. Bedford Institute of Oceanography, Report B.I.O. 65–6. Unpublished manuscript.

LÉVY, A. 1971 Eaux saumâtres et milieux margino-littoraux. *Revue Géogr. phys. Géol. dyn.*, 13, 269–78.

LIDZ, L. 1966 Planktonic Foraminifera in the water column of the mainland shelf off Newport Beach, California. *Limnol. Oceanogr.*, 11, 257–63.

LIPPS, J. H. and VALENTINE, J. W. 1970 The rôle of Foraminifera in the trophic structure of marine communities. *Lethaia*, 3, 279–86.

LISTER, J. J. 1895 Contributions to the life history of the Foraminifera. *Phil. Trans. R. Soc.*, B, 186, 401–53.

LOEBLICH, A. R. JR., and TAPPAN, H. 1964 Sarcodina, chiefly 'thecamoebians' and

Foraminiferida. In: Moore, R. C., Ed., *Treatise on invertebrate paleontology*. Geol. Soc. Amer., New York, pt. C, vols. 1–2, 900 pp.

LOGAN, B. W. 1969 Carbonate sediments and reefs, Yucatán Shelf, Mexico. Part 2. Coral reefs and banks, Yucatán Shelf, Mexico. *Mem. Am. Ass. Petrol. Geol.*, 11, 129–98.

LOGAN, B. W. and CEBULSKI, D. E. 1970 Sedimentary environments of Shark Bay, Western Australia. *Mem. Am. Ass. Petrol. Geol.*, 13, 1–37.

LOGAN, B. W., HARDING, J. L., AHR, W. M., WILLIAMS, J. D. and SNEAD, R. G. 1969 Carbonate sediments and reefs, Yucatán Shelf, Mexico. Part 2. Coral reefs and banks, Yucatán shelf, Mexico. *Mem. Am. Ass. Petrol. Geol.*, 11, 5–128.

LOOSE, T. 1970 Turbulent transport of benthonic Foraminifera. *Contr. Cushman Fdn foramin. Res.*, 21, 164–6.

LUDWICK, J. C. and WALTON, W. R. 1957 Shelf-edge, calcareous prominences in northeastern Gulf of Mexico. *Bull. Am. Ass. Petrol. Geol.*, 41, 2054–101.

LUKINA, T. G. 1967 The distribution of calcareous Foraminifera in the central Pacific. *Dokl. Akad. Nauk SSSR.*, 177, 1205–7. (Translation by A.G.I.)

LUTZE, G. F. 1964 Zum Färben rezenter Foraminiferen. *Meyniana*, 14, 43–7.

LUTZE, G. F. 1965 Zur Foraminiferen-Fauna der Ostsee. *Meyniana*, 15, 75–142.

LUTZE, G. F. 1968a Jahresgang der Foraminiferen-Fauna in der Bottsand Lagune (westliche Ostsee). *Meyniana*, 18, 13–30.

LUTZE, G. F. 1968b Siedlungs-Strukturen rezenter Foraminiferen. *Meyniana*, 18, 31–4.

LYNTS, G. W. 1962 Distribution of recent Foraminifera in Upper Florida Bay and associated sounds. *Contr. Cushman Fdn foramin. Res.*, 13, 127–44.

LYNTS, G. W. 1966 Variation of Foraminiferal standing crop over short lateral distances in Buttonwood Sound, Florida Bay. *Limnol. Oceanogr.*, 11, 562–6.

MACAROVICI, N. and CEHAN-IONESI, B. 1962 Distribution des Foraminifères sur la plate-forme continentale du nord-ouest de la Mer Noire. *Trav. Mus. Hist. nat. 'Gr. Antipa'*, 3, 45–60.

MCENERY, M. and LEE, J. J. 1970 Tracer studies on calcium and strontium mineralization and mineral cycling in two species of Foraminifera, *Rosalina leei* and *Spiroloculina hyalina*. *Limnol Oceanogr.*, 15, 173–82.

MCGLASSON, R. H. 1959 Foraminiferal biofacies around Santa Catalina Island, California. *Micropaleontology*, 5, 217–40.

MCMASTER, R. L. and GARRISON, L. E. 1966 Mineralogy and origin of southern New England shelf sediments. *J. sediment. Petrol.*, 36, 1131–42.

MADEIRA, M. 1969 Foraminifera from Sao Francisco do Sul, State of Santa Catarina, Brazil. *Iheringia*, Zool., 37, 3–29.

MAIKLEM, W. R. 1968 Some hydraulic properties of bioclastic carbonate grains. *Sedimentology*, 10, 101–9.

MARE, M. F. 1942 A study of a marine benthic community, with special reference to the microorganisms. *J. mar. biol. Ass. U.K.*, 25, 517–74.

MARSZALEK, D. S., WRIGHT, R. C. and HAY, W. W. 1969 Function of the test in Foraminifera. *Trans. Gulf-Cst Ass. geol. Socs*, 18, 341–52.

MATEU, G. 1968 Contribution al conocimiento de los Foraminiferos que sirven de alimento a la Holoturias. *Bol. Soc. Hist. nat. Baleares*, 14, 5–17.

MATOBA, Y. 1970 Distribution of recent shallow water Foraminifera of Matsushima Bay, Miyagi Prefecture, North-east Japan. *Sci. Rep. Tôhoku Univ.*, Ser 2, Geol., 42, 1–85.

MELLO, J. F. and BUZAS, M. 1968 An application of cluster analysis as a method of determining biofacies. *J. Paleont.*, 42, 747–58.

MENZIES, R. J. and GEORGE, R. Y. 1967 A re-evaluation of the concept of hadal or ultra-abyssal fauna. *Deep-Sea Res.*, 14, 703–22.

MILLER, R. L. and KHAN, J. S. 1962 *Statistical analysis in the geological sciences.* John Wiley and Sons, New York, 483 pp.

MOORE, W. E. 1957 Ecology of recent Foraminifera in northern Florida Keys. *Bull. Am. Ass. Petrol. Geol.*, 41, 727–41.

MORTON, J. E. 1959a The adaptations and relationships of the Xenophoridae (Mesogastropoda). *Proc. malac. Soc. Lond.*, 33, 89–101.

MORTON, J. E. 1959b The habits and feeding organs of *Dentalium entalis. J. mar. biol. Ass. U.K.*, 38, 225–38.

MOULINIER, M. 1967 Répartition des Foraminifères benthiques dans les sédiments de la Baie de Seine entre le Cotentin et le meridien de Ouistreham. *Cah. océanogr.*, 19, 477–94.

MULLER, W. A. and LEE, J. J. 1969 Apparent indispensibility of bacteria in foraminiferan nutrition. *J. Protozool.*, 16, 471–8.

MURRAY, J. 1895 A Summary of the scientific results – general observations on the distribution of marine organisms. *Rept. Voy. Challenger*, 1431–62.

MURRAY J. W., 1963 Ecological experiments on Foraminiferida. *J. mar. biol. Ass. U.K.*, 43, 631–42.

MURRAY, J. W. 1965a The Foraminiferida of the Persian Gulf. 2. The Abu Dhabi Region. *Palaeogeography, Palaeoclimatol., Palaeoecol.*, 1, 307–32.

MURRAY, J. W. 1965b Two species of British recent Foraminiferida. *Contr. Cushman Fdn foramin. Res.*, 16, 148–50.

MURRAY, J. W. 1965c On the Foraminiferida of the Plymouth Region. *J. mar. biol. Ass. U.K.*, 45, 481–505.

MURRAY, J. W. 1965d Significance of benthic foraminiferids in plankton samples. *J. Paleont.*, 39, 156–7.

MURRAY, J. W. 1966a The Foraminiferida of the Persian Gulf. 3. The Halat al Bahrani region. *Palaeogeography, Palaeoclimatol., Palaeoecol.*, 2, 59–68.

MURRAY, J. W. 1966b The Foraminiferida of the Persian Gulf. 4. Khor al Bazam. *Palaeogeography, Palaeoclimatol., Palaeoecol.*, 2, 153–69.

MURRAY, J. W. 1966c The Foraminiferida of the Persian Gulf. 5. The shelf off the Trucial Coast. *Palaeogeography, Palaeoclimatol., Palaeoecol.*, 2, 267–78.

MURRAY, J. W. 1967a Transparent and opaque foraminiferid tests. *J. Paleont.*, 41, 791.

MURRAY, J. W. 1967b Production in benthic foraminiferids. *J. nat. Hist.*, 1, 61–8.

MURRAY, J. W. 1968a The living Foraminiferida of Christchurch Harbour, England. *Micropaleontology*, 14, 83–96.

MURRAY, J. W. 1968b Living foraminifers of lagoons and estuaries. *Micropaleontology*, 14, 435–55.

MURRAY, J. W. 1969 Recent foraminfers from the Atlantic continental margin of the United States. *Micropaleontology*, 15, 401–9.

MURRAY, J. W. 1970a The Foraminiferida of the Persian Gulf. 6. Living forms in the Abu Dhabi region. *J. nat. Hist.*, 4, 55–67.

MURRAY, J. W. 1970b The Foraminiferida of the hypersaline Abu Dhabi Lagoon, Persian Gulf. *Lethaia*, 3, 51–68.

MURRAY, J. W. 1970c Foraminifers of the Western Approaches to the English Channel. *Micropaleontology*, 16, 471–85.

MURRAY, J. W. 1971a *An Atlas of British Recent Foraminiferids.* Heinemann, London, 244 pp.

MURRAY, J. W. 1971b Living foraminiferids of tidal marshes—a review. *J. Foramin Res.*, 1, 153–61.

MURRAY, J. W. and WRIGHT, C. A. 1970 Surface textures of calcareous foraminiferids. *Palaeontology*, 13, 184–7.

MYERS, E. H. 1935 Culture methods for the marine Foraminifera of the littoral zone. *Trans. Am. microsc. Soc.*, 54, 264–7.

MYERS, E. H. 1940 Observations on the origin and fate of flagellated gametes in multiple tests of *Discorbis* (Foraminifera). *J. mar. biol. Ass. U.K.*, 24, 201–26.

MYERS, E. H. 1941 Ecological studies of the Foraminifera. Nat. Res. Council, Division Geol. Geog., Ann. Report, 43–6.

MYERS, E. H. 1942a Ecologic relationships of some recent and fossil Foraminifera. Nat. Res. Council, Division Geol. Geogr., Ann. Report, 31–6.

MYERS, E. H. 1942b A quantitative study of the productivity of Foraminifera in the sea. *Proc. Am. phil. Soc.*, 85, 325–42.

MYERS, E. H. 1943a Life activities of Foraminifera in relation to marine ecology. *Proc. Am. phil. Soc.*, 86, 439–58.

MYERS, E. H. 1943b Ecologic relationships of larger Foraminifera. Nat. Res. Council, Division Geol. Geog., Ann. Report, Appendix Q, 26–30.

NICHOLS, M. M. and ELLISON, R. L. 1967 Sedimentary patterns of microfauna in a coastal plain estuary. In: Lauff, G. H., Estuaries, *Publs Am. Ass. Advmt Sci.*, 83, 283–8.

NICHOLS, M. M. and NORTON, W. 1969 Foraminiferal populations in a coastal plain estuary. *Palaeogeography, Palaeoclimatol., Palaeoecol.*, 6, 197–213.

NIKOLJUK, V. F. 1968 Was bergen die Erdschichten der Wüste Kara-Kum? *Pedobiologia*, 7, 335–52.

NIELSEN, E. S. 1963 Fertility of the oceans. Productivity, definition and measurement. In: Hill, M. N., ed. *The Sea*, 2, 129–64.

NYHOLM, K. G. 1957 Orientation and binding power of recent monothalamous Foraminifera in soft sediments. *Micropaleontology*, 3, 75–6.

NYHOLM, K. G. 1961 Morphogenesis and biology of the foraminifer *Cibicides lobatulus. Zool. Bidr. Upps.*, 33, 157–96.

NYHOLM, K. G. 1962 A study of the foraminifer *Gypsina. Zool. Bidr. Upps.*, 33, 201–6.

ODUM, H. T., CANTLON, J. E. and KORNICKER, L. S. 1960 An organizational hierarchy postulate for the interpretation of species-individual distributions, species entropy, ecosystem evolution, and the meaning of a species-variety index. *Ecology*, 41, 395–9.

ORTON, J. H. 1920 Sea-temperature, breeding and distribution of marine animals. *J. mar. biol. Ass. U.K.*, 12, 339–66.

ORTON, J. H. 1927 On the mode of feeding of the hermit-crab *Eupagurus bernhardus* and some other Decapoda. *J. mar. biol. Ass. U.K.*, 14, 909–21.

PANTIN, C. F. A. 1960 *Notes on microscopic technique for zoologists.* Cambridge University Press. 76 pp.

PARKER, F. L. 1948 Foraminifera of the continental shelf from the Gulf of Maine to Maryland. *Bull. Mus. comp. Zool. Harv.*, 100, 213–41.

PARKER, F. L. 1954 Distribution of the Foraminifera in the north-eastern Gulf of Mexico. *Bull. Mus. comp. Zool. Harv.*, 111, 452–588.

PARKER, F. L. and ATHEARN, W. D. 1959 Ecology of marsh foraminifera in Poponesset Bay. *J. Paleont.*, 33, 333–43.

PARKER, R. H. 1964 Zoogeography and ecology of macro-invertebrates of Gulf of California and continental slope of Western Mexico. *Mem. Am. Ass. Petrol. Geol.*, 3, 331–76.

PELTO, C. R. 1954 Mapping of multicomponent systems. *J. Geol.*, 62, 501–11.

PERATH, I. 1966 Living Foraminifera in the littoral zone of Achziv (northern Israel). *Israel J. earth Sci.*, **15**, 64–70.

PHLEGER, F. P. 1951 Ecology of Foraminifera, north-west Gulf of Mexico. Pt. 1. Foraminifera distribution. *Mem. geol. Soc. Am.*, **46**, 1–88.

PHLEGER, F. B. 1952 Foraminiferal ecology off Portsmouth, New Hampshire. *Bull. Mus. comp. Zool. Harv.*, **106**, 316–90.

PHLEGER, F. B. 1954 Ecology of Foraminifera and associated micro-organisms from Mississippi Sound and environs. *Bull. Am. Ass. Petrol. Geol.*, **38**, 584–647.

PHLEGER, F. B. 1955 Ecology of Foraminifera in south-eastern Mississippi Delta area. *Bull. Am. Ass. Petrol. Geol.*, **39**, 712–52.

PHLEGER, F. B. 1956 Significance of living Foraminiferal populations along the central Texas coast. *Contr. Cushman Fdn foramin. Res.*, **7**, 106–51.

PHLEGER, F. B. 1960a *Ecology and distribution of recent Foraminifera.* John Hopkins Press, Baltimore. 297 pp.

PHLEGER, F. B. 1960b Sedimentary patterns of microfaunas in northern Gulf of Mexico. In: Shepard, F. P., Phleger, F. B., and van. Andel, Tj. H., eds., *Recent sediments, North-west Gulf of Mexico*, Am. Ass. Petrol. Geol., 267–301.

PHLEGER, F. B. 1960c Foraminiferal populations in Laguna Madre, Texas. *Sci. Rep. Tôkoku Univ.*, Spec. vol. **4**, 83–91.

PHLEGER, F. B. 1964a Patterns of living benthonic Foraminifera, Gulf of California. *Mem. Am. Ass. Petrol. Geol.*, **3**, 377–94.

PHLEGER, F. B. 1964b Foraminiferal ecology and marine geology. *Mar. Geol.*, **1**, 16–43.

PHLEGER, F. B. 1965a Patterns of marsh Foraminifera, Galveston Bay, Texas. *Limnol. Oceanogr.*, **10**, Supplement, R 169–84.

PHLEGER, F. B. 1965b Sedimentology of Guerrero Negro Lagoon, Baja California, Mexico. In: Whittard, W. F. and Bradshaw, R., Submarine Geology and Geophysics, *Colston Pap.*, 205–35.

PHLEGER, F. B. 1965c Depth patterns of benthonic Foraminifera in the eastern Pacific. *Progress in Oceanography*, **3**, 273–87.

PHLEGER, F. B. 1966a Patterns of living marsh Foraminifera in south Texas coastal lagoons. *Boln Soc. geol. mex.*, **28**, 1–44.

PHLEGER, F. B. 1966b Living Foraminifera from coastal marsh, south-western Florida. *Boln Soc. geol. mex.*, **28**, 45–60.

PHLEGER, F. B. 1967 Marsh foraminiferal patterns, Pacific Coast of North America. *Cienc. del. Mar y Limm. Mex.*, 11–38.

PHLEGER, F. B. 1969 Some general features of coastal lagoons. *Mem. Simp. Intern. Lagunas Costeras*, UNAM-UNESCO, 5–26.

PHLEGER, F. B. 1970 Foraminiferal populations and marine marsh processes. *Limnol. Oceanogr.*, **15**, 522–34.

PHLEGER, F. B. and BRADSHAW, J. S. 1966 Sedimentary environments in a marine marsh. *Science, N.Y.*, **154**, (3756) 1551–3.

PHLEGER, F. B. and EWING, G. C. 1962 Sedimentology and oceanography of coastal lagoons in Baja California, Mexico. *Bull. geol. Soc. Am.*, **73**, 145–82.

PHLEGER, F. B. and LANKFORD, R. R. 1957 Seasonal occurrences of living benthonic foraminifera in some Texas Bays. *Contr. Cushman Fdn foramin. Res.*, **8**, 93–105.

PROVASOLI, L., MCLAUGHLIN, J. J. A. and DROOP, M. R. 1957 Development of artificial media for marine algae. *Arch. Mikrobiol.*, **25**, 292–428.

RAINWATER, E. H. 1966 The geological importance of deltas. In: Shirley, M. L. and Ragsdale, J. A. *Deltas in their geologic framework*. Houston Geological Society, U.S.A.

REITER, M. 1959 Seasonal variations in intertidal Foraminifera of Santa Monica Bay, California. *J. Paleont.*, 33, 606–30.

RESIG, J. 1958 Ecology of Foraminifera in the Santa Cruz Basin, California. *Micropaleontology*, 4, 287–308.

RICHTER, G. 1964a Zur Ökologie der Foraminiferen I Die Foraminiferen-Gesellschaften des Jadegebietes. *Natur. Mus., Frankf.*, 94, 343–53.

RICHTER, G. 1964b Zur Ökologie der Foraminiferen II Lebensraum und Lebensweise von *Nonion depressulum, Elphidium excavatum* und *Elphidium selsyense. Natur. Mus., Frankf.*, 9, 421–30.

RICHTER, G. 1965 Zur Ökologie der Foraminiferen III Verdriftung und Transport in der Gezeitenzone. *Natur. Mus., Frankf.*, 95, 51–62.

RICHTER, G. 1967 Faziesbereiche rezenter und subrezenter Wattensedimente nach ihren Foraminiferen-Gemeinschaften. *Senckenberg. leth.*, 48, 291–335.

ROBINSON, R. A. 1954 The vapour pressure and osmotic equivalence of sea water. *J. mar. biol. Ass. U.K.*, 33, 449–55.

RODEN, G. I. 1964 Oceanographic aspects of Gulf of California. *Mem. Am. Ass. Petrol. Geol.*, 3, 30–58.

ROUVILLOIS, A. 1970 Biocoenose et taphrocoenose de Foraminifères sur le plateau continental atlantique au large de l'île d'Yeu. *Revue Micropaléont.*, 13, 188–204.

RUSNAK, G. A. 1960 Sediments of Laguna Madre. In: *Recent Sediments, northwest Gulf of Mexico.* Tulsa: Amer. Assoc. Petr. Geol. 153–96.

RUSNAK, G. A., FISHER, R. L. and SHEPARD, F. P. 1964 Bathymetry and faults of Gulf of California. *Mem. Am. Ass. Petrol. Geol.*, 3, 59–75.

SAIDOVA, KH. M. 1967a The biomass and quantitative distribution of live Foraminifera in the Kurile–Kamchatka Trench area. *Dokl. Akad. Nauk SSSR*, 174, 207–9. (Translation by A.G.I.)

SAIDOVA, KH. M. 1967b Sediment stratigraphy and paleogeography of the Pacific Ocean by benthonic Foraminifera during the Quaternary. *Progress in Oceanography*, 4, 143–51.

SANDERS, H. L. 1960 Benthic studies in Buzzards Bay III The structure of the soft-bottom community. *Limnol. Oceanogr.*, 5, 138–53.

SANDERS, H. L. 1968 Marine benthic diversity: a comparative study. *Am. Nat.*, 102, 243–82.

SANDERS, H. L. 1969 Benthic marine diversity and the stability-time hypothesis. Diversity and stability in ecological systems. Brookhaven Symposia in Biology, no. 22, 71–80.

SANDON, H. 1932 The food of Protozoa. *Publs Fac. Sci. Egypt. Univ.*, 1, 1–187.

SCHAFER, C. T. 1969 Distribution of Foraminifera along the west coast of Hudson and James Bays, a preliminary report. *Maritime Sediments*, 5, 90–4.

SCHAFER, C. T. 1970 Studies of benthonic Foraminifera in Restigouche Estuary: 1 Faunal distribution patterns near pollution sources. *Maritime Sediments*, 6, 121–34.

SCHAFER, C. T. and SEN GUPTA, B. K. 1968 Benthic foraminiferal ecology in Port Castries Bay, St. Lucia: a preliminary report. *Maritime Sediments*, 4, 57–63.

SCHAFER, C. T. and PRAKASH, A. 1968 Current transport and deposition of foraminiferal tests, planktonic organisms and lithogenic particles in Bedford Basin, Nova Scotia. *Maritime Sediments*, 4, 100–3.

SCHEFFEN, W. 1940 Skeleton-features of large Foraminifera (especially Lepidocyclinidae) and their variation during life and fossilization. *Natuurk. Tijdschr. Ned.-Indie*, 100, 146–73.

SCHNITKER, D. 1967 Variation in test morphology of *Triloculina linneiana* d'Orbigny in laboratory cultures. *Contr. Cushman Fdn foramin. Res.*, 18, 84–6.

SCHOTT, W. 1935 Die Foraminiferen in dem äquatorialen Teil des Atlantischen Ozeans. *Deutsche Atlantische Exped.*, 6, 411–616.

SCHREIBER, E. 1927 Die Reinkultur von marinen Phytoplankton und deren Bedeutung fur die Erforschung der Produktionsfähigkeit des Meerwassers. *Wiss. Meeresunters.*, Abt. Helgoland, 10, 1–34.

SCHULTZE, M. S. 1854 *Ueber den Organismus der Polythalamien (Foraminiferen) nebst Bemerkungen über die Rhizopoden im Allgemeinen.* Angelmann, Leipzig.

SEGERSTRÅLE, S. G. 1957 Baltic Sea. *Mem. geol. Soc. Am.*, 67, (1), 751–800.

SEGURA, L. R. 1963 Sistematica y distribucion de los Foraminiferos litorales de la 'Playa Washington', al sureste de Matamoros, Tamaulipas, Mexico. *Boln. Inst. Geol. Méx.*, 68, 1–92.

SEIGLIE, G. A. 1970 The distribution of the foraminifers in Yabucoa Bay, southeastern Puerto Rico and its paleoecological significance. *Revista Española de Micropaleontologia*, 2, 183–208.

SEIGLIE, G. A. 1971 Distribution of foraminifers in the Cabo Rojo Platform and their paleoecological significance. *Revista Española de Micropaleontologia*, 3, 5–33.

SEN GUPTA, B. K. 1971 The benthonic Foraminifera of the Tail of the Grand Banks. *Micropaleontology*, 17, 69–98.

SEN GUPTA, B. K. and MCMULLEN, R. M. 1969 Foraminiferal distribution and sedimentary facies on the Grand Banks of Newfoundland. *Can. J. Earth Sci.*, 6, 475–87.

SHAFFER, B. L. 1965 A measure of community organization and ecosystem maturity in the fossil record. *J. Paleont.*, 39, 281–3.

SHEEHAN, R. and BANNER, F. T. 1972 The pseudopodia of *Elphidium incertum*. *Revista Española de Micropaleontologia*, 4, 31–63.

SHEPARD, F. P. and COHEE, G. V. 1936 Continental shelf sediments off the Mid-Atlantic states. *Bull. geol. Soc. Am.*, 47, 441–57.

SHEPARD, F. P. and EMERY, K. O. 1941 Submarine topography off California coast: Canyons and Tectonic interpretations. *Spec. Publs geol. Soc. Am.*, 31, 1–171.

SHIFFLET, E. 1961 Living, dead, and total foraminiferal faunas, Heald Bank, Gulf of Mexico. *Micropaleontology*, 7, 45–54.

SIDDALL, J. D. 1878 The Foraminifera of the River Dee. *Proc. Chester Soc. nat. Sci.*, 2, 42–56.

SIMPSON, E. H. 1949 Measurement of diversity. *Nature*, 163, 688.

SLITER, W. V. 1965 Laboratory experiments on the life cycle and ecological controls of *Rosalina globularis* d'Orbigny. *J. Protozool.*, 12, 210–15.

SLITER, W. V. 1971. Predation on benthic foraminifers. *J. Foramin. Res.*, 1, 20–9.

SLOBODKIN, L. B. and SANDERS, H. L. 1969 On the contribution of environmental predictability to species diversity. Diversity and stability in ecological systems, Brookhaven Symposia in Biology, no. 22, 82–93.

SMITH, P. B. 1963 Recent Foraminifera off Central America. Quantitative and qualitative analysis of Family Bolivinidae. *Prof. Pap. U.S. geol. Surv.*, 429-A, 1–39.

SMITH, P. B. 1964 Recent Foraminifera off Central America. Ecology of benthic species. *Prof. Pap. U.S. geol. Surv.*, 429-B, 1–55.

SMITH, R. K. 1968 An intertidal *Marginopora* colony in Suva Habor, Fiji. *Contr. Cushman Fdn foramin. Res.*, 14, 12–17.

SUGDEN, W. 1963 The hydrology of the Persian Gulf and its significance in respect to evaporate deposition. *Am. J. Sci.*, 261, 741–55.

SVERDRUP, H. U., JOHNSON, M. W. and FLEMING, R. H. 1942 *The oceans, their physics, chemistry and general biology.* Prentice Hall, Inc., New York, 1087 pp.

TAPLEY, S. 1969 Foraminiferal analysis of the Miramichi Estuary. *Maritime Sediments*, 5, 30–9.

THORSON, G. 1957 Bottom communities (sublittoral or shallow shelf). *Mem. geol. Soc. Am.*, 67, (1), 461–534.

TIETJEN, J. H. 1971 Ecology and distribution of deep-sea meiobenthos off North Carolina. *Deep-Sea Res.*, 18, 941–57.

TODD, R. 1961 Foraminifera from Onotoa Atoll, Gilbert Islands. *Prof. Pap. U.S. geol. Surv.*, 354-H, 171–91.

TODD, R. 1965 A new *Rosalina* parasitic on a bivalve. *Deep-Sea Res.*, 12, 831–7.

TODD, R. and LOW, D. 1961 Nearshore Foraminifera of Martha's Vineyard Island, Massachusetts. *Contr. Cushman Fdn foramin. Res.*, 12, 5–21.

UCHIO, T. 1960 Ecology of living benthonic Foraminifera from the San Diego, California, area. *Spec. Publs Cushman Fdn*, 5, 1–72.

UCHIO, T. 1967 Foraminiferal assemblages in the vicinity of the Seto Marine Biological Laboratory, Shirahama-Cho, Wakayama-Ken, Japan (Part 1). *Publs Seto mar. biol. Lab.*, 15, 399–417.

UJIÍE, H. 1962 Introduction to statistical foraminiferal zonation. *J. geol. Soc. Japan*, 68, 431–50.

UJIÍE, H. and KUSUKAWA, T. 1969 Analysis of foraminiferal assemblages from Miyako and Yamada Bays, north-eastern Japan. *Bull. natn Sci. Mus. Tokyo*, 12, 735–72.

VOORTHUYSEN, J. H. VAN 1960 Die Foraminiferen des Dollart-Ems-Estuarium. *Verh. K. ned. geol.-mijnb. Genoot, Geol. Ser*, 19, 237–69.

WALTON, W. R. 1952 Techniques for recognition of living Foraminifera. *Contr. Cushman Fdn foramin. Res.*, 3, 56–60.

WALTON, W. R. 1955 Ecology of living benthonic Foraminifera, Todos Santos Bay, Baja California. *J. Paleont.*, 29, 952–1018.

WALTON, W. R. 1964a Recent foraminiferal ecology and paleoecology. In: Imbrie, J. and Newell, N. D., *Approaches to paleoecology*, John Wiley and Sons, New York. 151–237.

WALTON, W. R. 1964b Ecology of benthonic Foraminifera in the Tampa-Sarasota Bay area, Florida. In: Miller, R. L., Ed., *Papers in Marine Geology*, 429–54. Macmillan Co., New York.

WATKINS, J. G. 1961 Foraminiferal ecology around the Orange County, California, ocean sewer outfall. *Micropaleontology*, 7, 199–206.

WELANDER, P. 1969 Effects of planetary topography on the deep-sea circulation. *Deep Sea Res.*, Suppl. 16, 369–91.

WELLS, J. W. 1957 Coral reefs. *Mem. geol. Soc. Am.*, 67, (1), 609–31.

WILCOXON, J. A. 1964 Distribution of Foraminifera off the southern Atlantic coast of the United States. *Contr. Cushman Fdn foramin. Res.*, 15, 1–24.

WILLIAMS, C. B. 1964 *Patterns in the balance of nature*. Academic Press, London, 324 pp.

WINTER, F. W. 1907 Zur kenntniss der Thalamophoren 1. Untersuchung über *Peneroplis pertusus* (Forskål). *Arch. Protistenk.*, 10, 1–113.

WRIGHT, C. A. and MURRAY, J. W. 1972 Comparisons of modern and Palaeogene foraminiferid distributions and their environmental implications. *Mém. B.R.G.M.* 79, 87–96.

WRIGHT, R. 1968 Miliolidae (Foraminiferos) recientes del estuario del Rio Quequen Grande (Provincia de Buenos Aires). *Revta Mus. argent. Cienc. nat. Bernardino Rivadavia, Hidrobiol.*, 2, 225–56.

YULE, G. U. 1944 *The statistical study of literary vocabulary*. Cambridge Univ. Press, 306 pp.

ZALESNY, E. R. 1959 Foraminiferal ecology of Santa Monica Bay, California. *Micropaleontology*, 5, 101–26.

Index of Genera and Species of Foraminiferids

Bold type indicates summary of ecological data in Appendix 2.

General Index